D1165413

THE COMING
CONVERGENCE

ADVANCE PRAISE

"Stanley Schmidt is our advance scout, journeying ahead to the glorious, complex future that awaits us all, and he reports back in this fabulous book, chock-full of the same kind of lucid and insightful commentary that has made his *Analog* editorials must-reading for three decades now. Schmidt gives us the same kind of clearheaded thinking and cleanly written prose that we associated with Asimov and Sagan."

Robert J. Sawyer
Hugo Award–winning author of *Hominids*

"James Burke's *Connections* traces the ways in which diverse technologies tumble together to form the world of today. Here, Stan Schmidt shows how comparably diverse technologies may intertwine to form the world of tomorrow."

Marc Stiegler
Hugo Award finalist for *Valentina*

"A cogent, thoughtful primer on where our technology comes from, how it works, and where it might be heading."

Richard A. Lovett, JD, PhD
Science journalist and four-time AnLab Award winner

"Technological change is not slowing down, and Stan Schmidt deftly outlines the promises—and perils—to come. As has so often been the case, today's headline issues are not the most important!"

Stephen L. Gillett, PhD
Nanotechnology researcher and writer

STANLEY SCHMIDT

THE COMING CONVERGENCE

Surprising Ways Diverse Technologies Interact
to Shape Our World and Change the Future

 Prometheus Books

59 John Glenn Drive
Amherst, New York 14228–2119

Published 2008 by Prometheus Books

Inquiries should be addressed to
Prometheus Books
59 John Glenn Drive
Amherst, New York 14228–2119
VOICE: 716–691–0133, ext. 210
FAX: 716–691–0137
WWW.PROMETHEUSBOOKS.COM

12 11 10 09 08   5 4 3 2 1

Library of Congress Cataloging-in-Publication Data

Schmidt, Stanley.
   The coming convergence : surprising ways diverse technologies interact to shape our world and change the future / Stanley Schmidt ; foreword by Paul Levinson.
      p. cm.
   Includes bibliographical references and index.
   ISBN 978–1–59102–613–6 (hbk. : alk. paper)
   1. Technology—Forecasting. 2. Technology—Social aspects. 3. Biotechnology—Social aspects. 4. Information technology—Social aspects. I. Title.

T14.5.S316 2008
303.48'3—dc22

2008005596

Printed in the United States of America on acid-free paper

To Otto and Georgia Schmidt,
my links to the past
who started me thinking
about the future

CONTENTS

8   **CONTENTS**

FOREWORD

We've been aware of the profound impact of technology for a long time—at least as far back as the ancient Greeks and Romans, and Cicero, who observed in his *de Natura Deorum* how, "by the use of our hands, we bring into being within the realm of Nature, a second nature for ourselves."

But the unexpected consequences of technology—how a device designed to do A leads to B, or maybe A and B and C—has been a line of inquiry and realization far more rare. Siegfried Giedion made a signal contribution in his 1949 *Mechanization Takes Command*. In it he explored, for example, how the industrialization of bread making—outside the home—liberated women and had an enormous impact on family life. Harold Innis and Marshall McLuhan offered a series of penetrating observations on the unexpected consequences of media—the relationships between the alphabet and democracy, for instance—and I chipped in with a few in my 1997 *The Soft Edge: A Natural History and Future of the Information Revolution*. I looked there at how the printing press made Columbus's voyage a bankable reality in Europe, by putting word of it in everyone's hands and minds. The Vikings got here first, of course, around 1000 CE, but no one took word of their discoveries too seriously, because it was merely spoken, not printed.

None of the above explorers of unintended technological consequences, however, including me, approached such issues as a scientist. Stan Schmidt, who is not only a scientist but a science fiction writer and editor, picks up the ball in this fascinating book, with a fine array of compelling examples.

What do identification and treatment of brain tumors and September 11, 2001, have in common? Both were made possible by the convergence of initially unrelated technology—the combination of x-rays and computer processing of large amounts of data in the case of medical detection technology; the intersection of aviation and skyscraper construction (which, as I always like to point out, was in part made possible by the elevator) in the case of the destruction of the Twin Towers. The medical treatment was of course deliberate and intended, but the devastating terrorist attack was obviously not what aviation and big-building pioneers had in mind.

Schmidt explores this and many other results of technological convergence, with a keen eye as to how they cannot only be unexpected, but also sometimes dangerous and damaging. This lays bare one of the central realities of our human relationship with technology: although consequences of technologies, including convergences, can and are often unexpected, we humans still maintain the ultimate control over how they are applied. We can use them for good, or for evil. Unintended consequences do not let us off the moral hook.

Years ago, I gave a talk entitled "Guns, Knives, and Pillows," and I still find this analysis helpful in highlighting the control we have over our technologies. We tend to think that technologies have inherent tendencies toward good or evil uses, and they generally do. Guns tend to be used for bad things, pillows for good things, knives for either good or bad—to cut food or cut people. But on closer consideration, we find that a gun can be used to hunt food (good) and a pillow can be used to murder someone by suffocation (bad)—indeed, guns and pillows are really just special kinds of knives, as are CAT scans, airplanes, tall buildings, and all technologies. Medicine is invented to do good, but is the basis of germ warfare.

Splitting the atom, as it used to be called, led to destruction of cities, but can also generate energy for the home and help with reduction of greenhouse emissions.

Every student of technology and its surprising roots and consequences has favorite examples. Maybe one of the reasons I so enjoyed the book before you is Stan Schmidt's choices are a lot like mine—his chapter on the loom and the internet, for example, really strummed a chord, since one of my Master of Arts students, Gail S. Thomas, years ago did her thesis on "The Loom and the Keyboard." But I suspect that whatever your technological proclivities and favorites, you'll find *The Coming Convergence* a most satisfying intellectual adventure—a vividly guided tour by an expert about how those convergences of the past have helped create our present, and, even more intriguingly, the increasing role the author expects them to play in the future. It's a pleasure to welcome this book into the canon of technological inquiry.

Paul Levinson
New York City
November 2007

ACKNOWLEDGMENTS

During the preparation of this book I have been repeatedly impressed by the generosity and helpfulness of both friends and strangers, including Nick Ruggiero, Carmen Perrone, Abby Browning, Janis Ian, Sheila Williams, Brian Bieniowski, John Gunnison, Terry Gibbons, Richard K. Lyon, Michael Perrone, Cliff Pickover, Jenny Hunter, K. Eric Drexler, Rosa Wang, and Robert A. Freitas Jr. My thanks to all of them, and any others I may have forgotten to mention.

Special thanks for contributions "above and beyond" to my agent, Eleanor Wood; my editor, Linda Regan, and her colleagues at Prometheus Books; Dr. Henry G. Stratmann; my parents, Otto and Georgia Schmidt; and most especially (as always), my wife, Joyce.

INTRODUCTION
CONVERGING CURRENTS
Then, Now, and Tomorrow

What do these two events have in common?

I

Kimberly, a young woman who has recently married and sees herself as well launched toward a bright future, suddenly begins suffering severe headaches. Her doctor, finding no obvious cause, sends her to a specialist, who in turn sends her for a "CAT scan." This is a test in which her head is positioned and held steady for several minutes inside a large machine while, as she sees it, nothing of any clear significance is happening. She is sent home, still as frightened and unsure as before.

A few days later her doctor calls. The CAT scan has been analyzed and a tumor has been found deep in her brain—but the doctors now know exactly where it is, how big it is, and what shape it is. They don't know whether it's malignant. If it is, it could kill her—and soon, if nothing is done about it.

But, thanks to the knowledge obtained from the CAT scan, they can target the tumor so precisely that they can eliminate it with very little damage to the surrounding tissue. The process is not fun, but a couple of months later she is pain-free and back on the road to that rosy future.

II

On the clear, sunny morning of September 11, 2001, a jetliner bound from Boston to Los Angeles swerves off its planned course and turns toward New York City. A few minutes later, it smashes into the North Tower of the World Trade Center, killing everyone on board and a great many in the tower. A few minutes after that, another plane similarly strikes the South Tower, and it becomes clear that the city—and through it, the country— has become the target of a massive terrorist attack carried out by a handful of individuals. The upper parts of both structures are engulfed by flames as their occupants frantically struggle to get out. Within a couple of hours, to the horror of watchers all over the country, both towers collapse. Thousands are killed, many more injured, and far more lives—and the economy of a nation—are massively disrupted. The entire mood of the country has been transformed in a fraction of a morning, in ways that will affect everyone and last a long, long time. As I write this, a few years later, it's still far too early to tell just how profound, far-reaching, and permanent the effects will be.

And all because of the actions of a few fanatics.

At first glance, these two incidents might seem as different and unrelated as two occurrences could be. One (see plate 1) is a tale of newfound hope for an individual, with no large-scale significance except that it's representative of many such incidents, now happening so routinely that we tend to take them for granted. We forget how remarkable it is that so many people can now be saved who would have been written off as hopeless just a few decades ago. The other (see plate 2) is a real-life horror story: thousands of lives destroyed and millions disrupted, in little more than an instant, by a few individuals not only willing to die for their beliefs, but also to kill thousands of innocents for them—and, unfortunately, with the physical means to do so.

The common denominator is that both the lifesaving tool called the *CAT scan*, and the swift murder of thousands by a few individuals, were made possible by the convergence of two or more seemingly separate technologies. The CAT scan (short for "computerized axial tomography," and now usually further shortened to CT scan) is the result of intentionally bringing together the fields of x-ray imaging and electronic computation that can analyze complex relationships among vast amounts of data in a reasonable amount of time. The World Trade Center disaster resulted from the technologies of aviation and large-scale building coming together in ways that the inventors of neither had in mind.

These convergences and others like them will continue to produce radical developments that will force us to make difficult, unprecedented choices. Many of those will have life-or-death significance, not just for individuals, but for civilization itself.

For example:

Should we do everything we can to increase human longevity? If we do, the worldwide problems already caused by population growth will be made even worse. Will longer life spans mean we have to do something drastic to reduce birth rates? Are we willing and able to do that?

Should we continue to build very tall buildings, or do they represent too much of a liability? How much freedom and privacy are we willing to give up to be safer from terrorists? Is privacy itself still a tenable ideal, or an outworn relic of an extinct past?

Can we get the advantages of large-scale power and communication networks without making ourselves vulnerable to catastrophic breakdowns caused by a single technical glitch or act of sabotage? Are very large cities still necessary, or even viable, as a basis for large-scale civilization? If not, what can replace them, and how can we get there from here? If individuals can "grow" whatever they need, that will eliminate their dependence on such large infrastructures, but it will also destroy the basis of our whole economic system. How can we ease the transition to a better one?

Under what conditions, if any, should we allow human cloning? Should we "preserve" our dearly departed as artificial intelligences that can simulate the personality (if not the physical form) of the deceased and continue to interact with the living? If so, should such "artificial citizens," being in a sense continuations of "real" citizens, be allowed to vote?

To some people, the idea that we will actually face such choices may seem too fantastic, bizarre, or farfetched to take seriously. Yet we have already had to confront such problems as the ethics of organ transplants (should *faces* be transplanted?) and abortions (at what point and to what extent should a fetus be considered a human being?). We are only beginning to create a body of law to deal with thorny dilemmas arising on the electronic frontier—problems like identity theft and what copyright means in a world where copying has become trivially easy and cheap. All of those would have seemed just as farfetched a few short decades ago.

What we have seen so far is just the beginning. As technologies continue to converge, they will continue to produce new possibilities both exhilarating and horrifying. We will have to make choices to embrace the opportunities while avoiding the horrors.

The goal of this book is to think about how we can make such choices intelligently. The first step in doing this is to understand how converging technologies can lead to results that could never be anticipated by considering a single field in isolation. As a preview of how it works, let's take a quick look at the broad outlines of what happened in my opening examples.

MEDICINE, X-RAYS, AND COMPUTERS

I won't go into much detail about how a CT scan works just yet. For now, I will merely observe that three things were going on more or less concurrently in the nineteenth and the early and mid-twentieth centuries:

(1) Doctors were, as they had been for a very long time, trying to keep patients in good health by preventing disease and injuries and making repairs when something went wrong. A major problem they faced was that doing their job often required knowing what was going on deep inside the human body, and that was usually out of sight.

(2) In 1895 a German physicist named Wilhelm Conrad Röntgen, while looking for something else, serendipitously discovered a new kind of radiation which soon came to be known as x-rays. Those turned out to

be closely related to visible light, but were not visible to the eye. They did have one most intriguing new property, however. They passed easily through many materials that were opaque to visible light (such as skin and muscle), but were stopped or attenuated by other materials (such as bone or metal). Since those rays could be used to expose photographic film, they quickly became a diagnostic tool for medical doctors. If x-rays were passed through a patient's body to a piece of film, parts of the film would be darkened more or less depending on how much of the radiation got through. This depended in turn on how much of what kinds of tissue or foreign matter it had to traverse. Thus the film formed a picture of the inside of the body, letting a doctor or dentist see such things as the exact shape and nature of a tumor, fracture, or cavity, or the location of a bullet.

(3) In the early twentieth century, several researchers developed the first digital computers, machines that could do complicated mathematical calculations by some combination of automatic electrical and mechanical processes. The first working models used electromechanical switching devices called *relays* and were huge, slow, and of limited ability—quite "clunky" by today's standards. But in the ensuing decades, workers found ways to dispense with macroscopic moving parts and do similar operations electronically. First they used vacuum tubes (now largely forgotten), followed by transistors and, later, integrated circuits, which combined huge numbers of microscopic transistors on a single small chip. The result of those improving technologies was a steady, dramatic, and accelerating improvement in computing capabilities. Recent computers are far smaller, faster, and more powerful than those of past decades, which allows them to be used to do types of problems that were simply far too difficult before. Such as the CT scan, which is the result of the confluence of medicine, x-ray technology, and computing technology.

In principle, CT scans could have been done almost from the beginning. The basic idea is that instead of making a conventional x-ray—a flat picture of the body as seen from one angle—you use x-rays to construct a three-dimensional image of everything inside the body. All you have to do is shoot x-rays through the body in many different directions and measure how much comes out the other side in each case. Since different materials absorb x-rays more or less strongly, there's only one distribution of materials that would give the pattern of absorption that you measure. The problem is that figuring out what that distribution is requires solving many simultaneous equations for many unknown quantities. That can be done, but doing it with pencil and paper is extremely difficult and time-consuming. The patient would die of old age while waiting for the test results.

But if you have fast, powerful computers available, they can do just that sort of tedious "number crunching" quickly and very efficiently. Set up a machine to take a series of x-rays from different angles, program fast computers to solve those tangles of absorption equations in a short time, and you get one of the most powerful diagnostic tools in medicine.

AIRPLANES AND BIG BUILDINGS

In the World Trade Center incident, two technologies had developed independently, for quite different purposes. An incidental property of one of them—one that most sane people would try hard to avoid bringing into play—was used to attack an incidental vulnerability in the other.

Building very large structures isn't easy, but has a number of advantages if you can learn to do it. The original inspiration was the growth of crowded cities that needed lots of businesses to support their populations, but had relatively little land on which to put those businesses. Building up rather than out effectively increased that area by a large factor, allowing tens of thousands of people to work (or live) on a couple of blocks of land. Building high also posed a whole complex of related engineering challenges. Before structures like the Sears Tower or the World Trade Center could be built, engineers had to develop ways to support all that weight, to ventilate and heat the large volumes within, and to transport large numbers of people quickly and easily through large vertical distances. Once those problems were solved, the technology of building big was widely applied not only to bustling business centers but also to housing large populations.

Aviation, meanwhile, followed a path of its own, with a quite different, essentially simple goal: getting people or cargo from point A to point B. It's hard to say how far forward such pioneers as Orville and Wilbur Wright were looking, but they surely would have been astounded at some of the developments to which their early experiments would eventually lead. Quite possibly they saw it purely as a technical challenge, a problem to be solved "because it's there": to get a manmade, self-propelled machine off the ground and keep it there long enough to go somewhere else. Once that possibility was demonstrated, large potential advantages began to suggest themselves. An airplane would not require good roads—or any roads—all the way from its point of departure to its destination. All it would need would be a few airports, and in the beginning, those didn't have to be large or complicated. Since a plane en route would be above all the obstacles that a ground vehicle has to contend with—trees, buildings, hills, chasms, rivers—it could travel much faster.

Powered flight took a few years to catch on, but once it did, it attracted lots of talent and money to grow exponentially. Within a few decades big, fast airplanes had become one of the world's most important modes of transport for both people and goods. They also became weapons, at least indirectly. Fighter pilots machine gunned one another in World War I dogfights; and Orville Wright, the Ohio bicycle repairman who made that first brief flight at Kitty Hawk, lived to see planes drop atomic bombs on Japanese cities in World War II.

And in 2001, planes themselves became weapons. A plane big enough to carry hundreds of passengers, fully fueled for a transcontinental flight at close to the speed of sound, is itself a powerful bomb, if used in the wrong way.

And the same tall buildings that enabled thousands of people to work in a small area became tempting targets, concentrating thousands of potential victims in a compact, sharply defined bulls-eye.

RIVERS AND TRIBUTARIES OF CHANGE

Much of our past has been shaped by such convergences of what started out as independent lines of research or invention, and even more of our future will be shaped in this way. We live in what our ancestors, even quite recently, would have viewed as a "science fictional" age; but there's an important difference between our present reality and much science fiction. Many science fiction writers have tried to follow the principle, perhaps first enunciated by H. G. Wells, of limiting themselves to one contrary-to-present-knowledge assumption per story—for example, to use cases from Wells's own work, suppose someone developed a way to travel in time, or to make himself invisible. What has happened in reality (and is happening in more modern science fiction by writers seeking to imitate reality more believably) is that several new things develop concurrently, and important events grow out of their collision and interaction.

Try to look at my real-world examples from the viewpoint of a writer trying to imagine them before they happened—say, a writer working around 1870, shortly after the Civil War in the United States. Such a writer might have come up with a story in which everything was just as the United States was in 1870 *except* that somebody found a way to photograph the inside of the human body, *or* to build and fly airplanes, *or* to build huge buildings, *or* to make powerful computers that could be used for calculations too complicated for human bookkeepers. Such a story would follow Wells's advice diligently, but it would be a world little like the one we live in, and considerably less interesting. For the world we actually have at the beginning of the twenty-first century is the result of *all* of those things being done concurrently, by different people, and other people seeing ways to put them together to yield still other developments, even more surprising.

You could think of Wells's model of a story as following a single stream from its source, and real history as a vast map of a complicated river system, with streamlets coming from many sources, then flowing together to form bigger rivers, with branches occasionally splitting off and later recombining. (See plate 3.) The medical use of x-rays is a minor convergence, the confluence of the currents of diagnostic medicine and x-ray imaging. The high-speed computers that make CT scans possible result from another merging of two currents: those of digital computation (which can be done with a wide variety of switching devices) and semiconductor physics (which provides a way to make very *small* switching devices). The CT scan itself comes from the merger of those two larger currents—medical use of x-rays and high-speed semiconductor-based computers—each of which itself grew from the convergence of at least two smaller ones.

This kind of effect will continue to shape our future, quite likely in an ever-accelerating way. With ordinary streams, as more of them flow together, the total flow becomes greater and faster, at least until the combined flow carves out a deeper channel. The analogy isn't perfect, of course, but it does have at least a qualitative sort of validity. And when it comes to scientific and technological progress, there are additional effects contributing to that speedup.

In *Engines of Creation*, K. Eric Drexler's first book about nanotechnology (a new kind of molecular-scale technology), the author wrote about ever-increasing computer speed and power and their effect on human progress.[1] With exponentially increasing speed, computers made it possible to do, in weeks or days or minutes, work that had never been done before because it would have taken too many person-years of drudgery. The results of those calculations suggested new problems to be tackled; and often those could be solved even faster because while the first set was being done with last year's computers, this year's, an order of magnitude faster, were being developed and were now available for use.

And so on. Furthermore, whatever computers were currently available were being used not just by workers in field A, but also by others in fields B, C, and D. Sometimes somebody in one field would be clever enough to look across a boundary and see that a problem somebody had solved in a seemingly unrelated field could be applied to his own. To give an example that doesn't even depend on advanced computers, most people have at least some inkling of what a hologram is: a photographic image so truly three-dimensional that you can literally move your head and look behind objects in it. The basic principle was developed by Dennis Gabor in 1947, but not much happened with it for several years because holograms were very difficult to make and view with light sources available at that time. While holograms languished as an academic curiosity, another research current involving Alfred Kastler, Charles Hard Townes, and Theodore Harold Maiman led to the development of a new kind of light source: the laser. As far as I know, none of them was

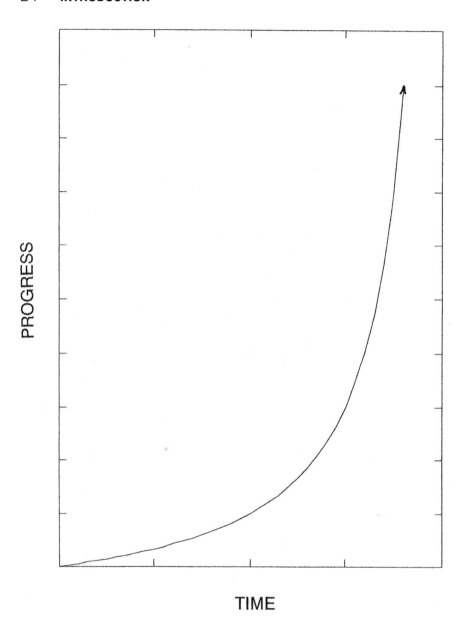

Figure 1. A singularity. The curve shown climbs ever more steeply, following the right margin ever more closely but never quite reaching it. (Diagram by the author.)

even thinking about holograms; but once the laser was available, it wasn't long before somebody noticed that it made holography relatively easy.[2] Holography then promptly took off as a hot field of research. The first wave of intensive holography research all used lasers, but then it wasn't long before people started looking for ways to make holograms that didn't depend on lasers. Now it's not uncommon to see holograms (of sorts) on magazine covers and credit cards.

Ever-accelerating computation and cross-fertilization among scientific disciplines can be expected to lead to dizzyingly rapid societal change. Science fiction writer Vernor Vinge, in his novel *The Peace War*,[3] imagined the cross-linked graphs of progress in all fields of endeavor becoming so steep that they became practically vertical, with change that would once have taken years occurring in minutes.

In mathematics, the place where a graph becomes vertical is called a *singularity*. Vinge, who is also a mathematician, called his fictitious time when graphs of historical change become so steep they're practically vertical *The Singularity*. That term has quickly come to be generally understood as one of the central concepts in contemporary science fiction.

But don't think it has no relevance to the real world. "Vinge's Singularity" is a phrase that was coined and came into wide use in a science fictional context, but Vinge was thinking very seriously about where we might really be headed. Eric Drexler, in his nonfictional chapters on accelerating change, was independently describing much the same thing, even if he wasn't using the same words. If you need an example to make you take such concepts seriously, let me mention one more forecast that Drexler made.

In chapter 14 of *Engines of Creation*, titled "The Network of Knowledge," Drexler described some of the limitations of information handling as it existed up to then, and envisioned a way to get beyond them.[4] At that time, most published information was scattered around in paper books, journals, and newspapers. It was hard to access whatever bit you might want, and easy to publish nonsense with little chance of anyone's ever seeing a retraction or rebuttal, even if one was published. Drexler's vision (which he called *hypertext*, a word coined by Theodor Nelson)[5] was of what amounted to a worldwide library, with the equivalent of thousands of libraries' content instantly available from any of thousands of connected computers. Instead of a footnote advising the reader of a related reference that might be found in some obscure journal, an article might include a "link" that would take an interested reader directly and immediately to that article to see for himself exactly what it said. Readers could post their reactions right with the article, so that anyone reading the article could also see everything anyone else had been interested enough to say about it. If the author changed his mind about something he'd written, he could post a correction right there where it couldn't be missed.

What he was describing was, in other words, something very much like the present-day internet*—but to most people in 1986, it still sounded pretty fantastic. Yet it's already here, in such a highly developed and ubiquitous form that almost everyone now takes it for granted. And it has contributed markedly to a change in how research is done. Scientists no longer have to wait months to publish their work or read their colleagues'; they can search quickly and easily for anything that's been done in their field; and they can argue about theories and experimental results almost as easily as if they were in the same room. All of which contributes to a general speedup and makes the Singularity that much more plausible.

Drexler wasn't kidding about that—and he's not kidding about the rest of it, either.

ULTIMATE PROMISE, ULTIMATE THREAT, OR BOTH?

Although the potentials of nanotechnology, for instance, may sound a lot like magic, we already have living proof that its principle can work. In fact, we *are* living proof that it can work.

The basic idea of nanotechnology is that instead of making things by cutting or otherwise shaping bulk material with tools that we can see and handle, extremely tiny machines—molecule-sized—could build just about

---

*No general consensus has yet emerged on whether "internet" should be capitalized; in this book, I do not capitalize it because the internet is more analogous to "the power grid" or "the telephone system" than to "Consolidated Edison" or "Verizon."

anything by putting it together, atom by atom. We are a proof of that principle because that's essentially what happens in living organisms. Biology is a sort of natural nanotechnology, and its existence proves that this sort of thing can happen at least as efficiently as it does in biological systems. It does not rule out the possibility that it might be done in a much wider range of applications by using artificial nanomachines. If it can, it's quite possible that we will eventually live in a radically transformed society in which practically anything can be cheaply grown rather than expensively manufactured. When Eric Drexler was publishing his initial speculations on nanotechnology (or "molecular engineering") in the mid-eighties, many thought it sounded like the wildest sort of science fiction, but a great many real scientists are now actively working on it. I know several of them personally—including at least one who just a few years ago was a vocal skeptic, but is now quite busy *doing* nanotechnology.

Biology itself is, of course, another of the major currents of research that will influence our future. Everyone has read about the controversies over cloning, stem cell research, and genetic engineering. Some would prefer that those things just go away, because they raise the need for difficult, unfamiliar choices. But they won't go away; once the capabilities exist, and the potential rewards in such areas as medicine are seen, those things *will* happen. The choices are about when and how.

Not surprisingly, our newfound computing abilities have played a major role in laying the groundwork for nanotechnology and biotechnology (biology-based technology). Much of our new understanding of biology and ability to manipulate it comes from the application of large-scale, high-speed computing to such problems as mapping the human genome. This is a problem that would have been far beyond the capabilities of pencil-and-paper biologists or mathematicians—not because it's so hard, but because it's so *big*.

We can foresee other powerfully influential convergences a bit farther down the road (or stream). If researchers do develop the atomically precise nanofactories that are currently seen as the ultimate goal of nanotechnology, they will need to be controlled by submicroscopic computers, which involves further development of the "computing" stream. Specialized nanomachines might also be developed for medical purposes, such as going into a living body to find and destroy cancer cells while leaving everything else alone.

Any powerful tool can do a great deal of good—or a great deal of harm. We have seen this repeatedly in the past with such developments as agriculture, airplanes, and nuclear energy. The tools that look likely to emerge from the convergence of research in computing, biology, and nanotechnology promise to be far more powerful than any we have known before. As such, they can transform our future lives in seemingly miraculous ways, or create nightmares almost beyond imagining. Can we learn to take advantage of the benefits while steering clear of the dangers? Maybe—but that ever-increasing speed of the rush of events, caused by faster and faster computers, faster and faster communication, and synergies between fields, means that those capabilities, with their attendant promises and perils, are likely to be upon us far sooner than we might expect. One early nanotechnology researcher, asked informally when we might expect a well-developed nanotechnology, said the optimistic estimate was thirty years; the pessimistic was ten.

The implication was not that nanotechnology is a bad thing that we can't avoid and would like to put off as long as possible. Rather, it's a potentially good thing that will involve such sweeping changes that we'll need time to prepare for them and figure out how to deal with them. And the changes that might be produced by nanotechnology converging with other fields such as biotechnology and information technology dwarf even those anticipated by considering nanotechnology alone.

Of course, it isn't really possible to consider nanotechnology alone. Making it work requires collaboration among materials scientists, information scientists, chemists, and physicists; and many of its potential applications involve other fields such as medicine. But nanotechnology itself was the most radical innovation in sight when Drexler wrote *Engines of Creation*. Anticipating the huge range of profound changes it could produce, and the speed with which those changes could happen, he and some colleagues established an organization called the Foresight Institute. The institute is dedicated to providing a clearinghouse for news about nanotechnology-related research and thinking about how we might best work toward reaping its rewards and avoiding its pitfalls.[6]

Toward the end of 2002, the National Science Foundation published a report and hosted a conference on "Converging Technologies for Improving Human Performance."[7] The specific areas they identified as major currents whose convergence would radically transform our future were a group identified by the unpronounceable acronym NBIC: Nanotechnology, Biotechnology, Information Technology, and Cognitive Science (the science of how we know, learn, and think). For the purposes of this book, we should keep in mind that those may not be the only important currents in even our near-term future, but they will almost certainly be among the most important, and they alone suggest astounding possibilities. The report foresees such things as direct connections between human brains and machines, tailored materials that adapt to changing environmental conditions, computers and environmental sensors integrated into everyday wear, and medical technologies that might eliminate such persistent problems as paralysis and blindness. Please bear in mind that this is not science fiction but was a recent serious attempt by scientists acting as such to foresee what might happen in the next few decades.

The report describes a "golden age" and a "new renaissance,"[8] but will such a future really be that, or an unprecedented kind of horror—or something in between, with elements of both? As my opening examples show, powerful technologies can be used for powerful benefits or great harm. In my novel *Argonaut*, a convergence of powerful computing with nanotechnology and another technology called *telepresence* enables very small numbers of individuals to gather and use vast amounts of information quickly.[9] This is fine if you're the one trying to learn a lot, but not nearly so good if someone else is using the same methods to gather information for use against you. Can we reap the benefits of the coming convergences without the dangers? If so, how? How can we learn to make the most intelligent choices?

Radical change will soon be upon us and we shall have to make crucial, difficult decisions. What I shall try to do in this book is:

(1) Describe some of the convergences that have led to our present world, tracing the development of some major past threads, including the stories of the people who made them happen, to see how they evolved from their beginnings to create pervasive parts of our present world.

(2) Describe some of the major lines of research that seem most likely to shape our (relatively near) future, beginning with the facts to date and going on to educated speculation about where current trends might lead.

(3) Examine how some of these current trends may interact to produce radically new abilities, and consider both the benefits and the dangers these might lead to.

(4) Consider how we might scout out our future options and steer the ship wisely. In a sense, this last is perhaps the most important. Making those synergies happen will require breaking down barriers between scientific and technological fields, by such means as interdisciplinary education aimed at making scientists and engineers comfortable with working across disciplinary boundaries and collaborating with colleagues in other fields. It will also pose political and economic challenges such as how to get the benefits to people and how to minimize economic disruption and the abuse of new abilities for antisocial purposes.

Since those changes will affect everybody, it is important for as many of us as possible to have some understanding of what may be coming. So here we go; and we might as well begin with the story of computing, since that has already had such a profound impact on so many areas of life and will continue to converge with and influence just about everything else.

CHAPTER 1
FROM FABRIC LOOMS TO THE INTERNET
The Story of Computing

A French textile mill at the end of the eighteenth century might seem an odd place to look for the origins of the computers that now pervade our lives, but metaphorical rivers, like real ones, tend to have many sources. Where does the Mississippi begin? The answer is usually said to be Lake Itasca in Minnesota. But I once flew over the spot where the Ohio flows into the Mississippi and the two rivers looked so evenly matched that I wondered why the Ohio is considered a tributary of the Mississippi instead of vice versa. It *could* be thought of that way; but the real answer is Lake Itasca *and* the several origins of the Ohio *and* the several sources of the Missouri.

And so it is with the history of invention and civilization. Nobody simply thought up the modern computer out of the blue, and French inventor Joseph-Marie Jacquard was surely not thinking of anything like modern supercomputers or the internet, or of computing at all. But his key invention turned out to be perhaps the earliest source of a current that led to those ends, and its real importance lay at least as much there as in what he originally had in mind.

Jacquard was born in July 1752, and in 1790 he hatched the idea for an automatic loom using punched cards to control the motions of the machine and thereby weave cloth in any desired pattern. The patterns could be quite intricate, and if you wanted a different one, you didn't have to build a new loom—just put in a different set of punched control cards.

As often happens with inventions, the path from inspiration to widespread application was not smooth or uninterrupted. Jacquard got sidetracked by the French Revolution and fought with the revolutionaries instead of continuing to work on his loom. Only afterward did he get back to it, and it wasn't until 1805 that he actually introduced the working machine that has come to be known as the Jacquard loom. As also often happens with inventions, the Jacquard loom was not welcomed universally with open arms.[1] Weavers who feared for their jobs burned Jacquard looms and attacked the inventor personally. But (as also often happens) the advantages of the new technology were so clear that it eventually came into general use anyway. By 1812, eleven thousand of the newfangled contraptions were in use in France; by the 1820s, it crossed the Channel into England; and from there, it spread throughout most of the world, the clear ancestor of the automatic looms now in wide use.

Meanwhile, across the Channel before the Jacquard loom got there, the English mathematician Charles Babbage was starting another channel of development: one that seems more obviously an ancestor of our modern computers. About the time the Jacquard loom was technologically revolutionizing French textile making, Babbage got the idea of mechanically calculating mathematical tables, and built a small prototype to demonstrate the feasibility of the principle.

The essence of mechanical computing is switching devices. For a simple example, in one of those little handheld counters, or a mechanical odometer for a car or bicycle, pressing a lever or turning the wheels a certain distance turns a wheel whose edge is marked with numbers representing "ones," or miles. When that wheel makes a full turn it catches a tooth on another wheel next to it, and turns that wheel a fraction of a turn, making a digit on its rim (representing "tens") advance by one. Other types of mechanical switches can embody "yes-no" logic: if the force applied to a spring-loaded lever is larger than a certain amount, it causes something to jump to a new position. If the applied force is too small, that doesn't happen. The applied force is the input; the output is 1 if it trips the switch and 0 if it doesn't. Several (or many) such switches may be connected together so that the output of one becomes the input for the next, determining whether it trips or doesn't.

Babbage was ambitious; he wanted to build machines that could do large numbers of high-precision calculations. To do that, he needed switches and mechanical linkages built to high standards of precision and reliability. He devoted much of his energy to learning to make them, with financial help from the British government and Cambridge University, where he was a professor from 1828 to 1839. The climax of his thinking was his analytical engine, for which he drew up detailed plans in the mid-1830s. This, had it actually been built, would have been an obvious ancestor of modern computers: incorporating devices for carrying out arithmetical operations according to the operator's instructions, storing numbers in mechanical memory, and controlling the sequence of operations.[2] His method for inputting the numbers to be processed, and the instructions for processing them, used punched cards much like those in the Jacquard loom.

A working model of Babbage's analytical engine was never completed because techniques for manufacturing metal machine parts with the required precision did not yet exist. So while Babbage made a promising start toward digital computers, he was not able to follow it through; it became, in the language of my "converging streams" analogy, a backwater, going nowhere. His work was forgotten until the late 1930s, but he was not the only person to notice the potential of Jacquard's punched-card system. Herman Hollerith, an American statistician, used something like it, and likely inspired by it, to automate the tabulation of US population data for the census of 1890—by which time sufficiently precise manufacturing techniques *were* available.

Computation, of course, is more than just tabulation; we need not only to *list* numbers, but also to calculate meaningful new ones from them. And as virtually everyone knows, computers are now employed for much more than numerical calculation: word processing, graphics, e-mail, and musical composition and scoring are just a few of its uses. Even if your main interest is number crunching, a problem of any complexity (and that's the kind for which computers shine) may require making choices and logical decisions as part of the process.

For example, royalties paid to the author of a book are commonly a percentage of sales, but if sales exceed a certain threshold, the percentage paid on additional sales is higher. So an accounting program to calculate royalties due might first add up the sales through all outlets. If that number is below the breakpoint, it will then calculate royalties using the lowest percentage. If it's higher, it will have to do that for copies up to the threshold, then determine how many additional copies were sold, calculate royalties on those using the higher percentage, and add the two figures.

Since most computer applications involve making many such choices, no description of the currents that converged to create computing as we know it can ignore the work of George Boole. The son of an English shoemaker, Boole gravitated early, with help and encouragement from his father, to an intense interest in mathematics and logic. At sixteen he was already teaching math in a private school, and by twenty he had started his own school. His great contribution was the observation of the close relationship between mathematics and logic, which had previously been regarded as quite separate. Boole developed a way of symbolically representing logical premises, relationships (like IF, AND, and OR), and conclusions in a way resembling the symbology of algebra. As in algebra, the symbols could be manipulated according to specified rules and give logically meaningful results. This Boolean algebra would later become the basis of computer switching procedures.

The physical devices used for those switching operations in early computers were, of course, at least partly mechanical. In the late 1930s and early 1940s several groups of researchers were working on what might be considered the earliest "modern" computers. Perhaps the best-known and historically most influential of those was the Harvard Mark I, or Automatic Sequence Controlled Calculator, built by Howard Aiken of Harvard University and a team of engineers at IBM. Like the Babbage and Hollerith machines, it used punched cards for data, and the switching devices it used for data processing were electromechanical *relays*.

Relays come in several types, but the basic idea is that a change in one circuit is used to cause a change in another. A simple and fairly familiar example, if you have a car of a certain age, is the starter solenoid. The starter motor in a car has to crank the whole engine, which is heavy, so it needs to handle a large electrical current, which means it has to *be* rather large. There isn't room to mount such a bulky device on the dashboard, and you probably wouldn't want a switch handling such strong currents right by your hand anyway, just in case something goes wrong. So the ignition switch you turn with your key is a little one that just turns on a small current. That current flows through an electromagnet that causes a plunger to move, closing the circuit that actually turns on the large current to the starter motor.

Even a dainty electromechanical relay is pretty big by today's standards, so it wasn't a very good long-term solution as the basic working element of a computer. The Harvard Mark I was huge by today's standards—some fifty feet long and eight feet high—yet what it could do was minimal compared to almost any of the laptops we see all around us today. So how did we get from there to here?

To answer that we have to look at a couple of other streams of research that might seem at first glance to have little to do with computers—because making computers significantly smaller, faster, and capable of storing and processing more data required new kinds of switching devices, fundamentally different from electromechanical relays.

The first of those "major tributaries" was electronics, in the original sense of devices involving free electrons moving in evacuated enclosures. Ironically, this had its roots in the one purely scientific discovery made by a man who was (and quite probably still is) the greatest inventor in human history.

Many laymen confuse science and invention, but while they're closely related and constantly interacting, they're not at all the same thing. Science is the attempt to understand how the universe works. Its major accomplishments are things like Newton's laws of motion, Maxwell's equations of electromagnetism, and Watson and Crick's explanation of how DNA transmits hereditary traits from one generation to another—elegant descriptions of general principles that describe how things work and make it possible to predict what will happen under new conditions. Invention, on the other hand, figures out ways to do things such as making useful alloys or flying machines. That process may involve conscious application of principles already formulated by science, but it doesn't have to. Inventors can and sometimes do hit upon useful techniques without understanding academically why they work.

Thomas Alva Edison, born in Milan, Ohio, in 1847, never claimed to be a scientist, but his accomplishments as an inventor seem almost superhuman.[3] He tossed revolutionary inventions off almost casually, beginning at a very early age. By the age of thirty he had established the first industrial research laboratory, employing some eighty scientists and dedicated expressly to churning out inventions. Of course, he would have denied the description of his work as "casual." Probably his most famous utterance was, "Genius is one percent inspiration and ninety-nine percent perspiration," and his whole working life was an illustration of that belief. Slow, analytical approaches were not his style. He read everything he could find that might relate to a problem and tried everything he could think of that might have a chance of solving it. Most of his attempts failed; but by working extraordinarily long hours punctuated by brief catnaps, he made so many of them that the small percentage that succeeded form a most impressive list, including the phonograph, motion pictures, and electric lighting.

The "Edison effect" (now usually called *thermionic emission*) was an almost accidental byproduct of his work on electric lights. The basic principle of the lightbulb is simple: a strong electric current is passed through a wire called the *filament*, heating it till it glows white. In practice, it was far from easy. First, the filament had to be enclosed in an evacuated bulb so it wouldn't simply burn up, and just getting a good enough vacuum took lots of work. Even after that, it took many more months of hard labor, with many trials and errors, to find filament materials and designs that could withstand the intense heat without melting or otherwise destroying themselves too quickly to be useful.

Even after Edison had a good enough light to patent and to introduce with a dramatic public demonstration on New Year's Eve of 1879, the experiments continued to refine the lightbulb's performance. In one experiment he sealed an additional metal plate into a lightbulb and was surprised to find electricity flowing across the vacuum gap from the glowing hot filament to the unheated plate. Edison himself didn't understand what was happening, but he did publish a description of the effect and patent the tube, even though he had no practical application in mind.

But John Ambrose Fleming, an English electrical engineer, did see an application for the Edison effect—a big one. Fleming, for a time in the 1880s, worked with Edison's London office on establishing the electric light industry in Britain. He made a closer study of what was happening in the "Edison effect bulb" and found that electrons were being "boiled" off the hot filament. If the extra metal plate was uncharged, the emitted electrons did little more than form a charged cloud around the filament; but if a positive charge was applied to the plate, the emitted electrons were attracted to it. If a complete circuit was made by wiring the plate to the positive terminal of a battery and the filament to the negative terminal, a current flowed through the circuit. If the battery was reversed, making the plate negative, emitted electrons were repelled back toward the filament from which they had come and no current flowed.

The real usefulness of that observation lay in a situation where no battery was involved. A battery drives current always in the same direction, a situation called *direct current*. For some purposes, such as power transmission over a network, alternating current—that is, current which cyclically reverses direction, which can be produced by a mechanical generator—works better. Fleming recognized that if an alternating current source was connected to an Edison effect tube, current would actually flow when and only when the plate was positive with respect to the filament; when the plate was negative, no current flowed. So the Edison effect tube, or diode (a tube with two electrodes), acted as a rectifier, turning alternating current (AC) into direct current (DC). Thus generators could send AC out over transmission lines, and then customers could use Fleming's rectifier to turn it into DC for applications that required it.

The ability to convert AC into DC was also useful in the new field of radio communication, which owed much to the Italian engineer Guglielmo Marconi.[4] Radio also inspired the Alabama-born inventor Lee De Forest to modify Fleming's diode in a simple way with enormous ramifications.[5] De Forest added a third element, called the *grid*, between filament and plate, making a triode (a vacuum tube having three elements instead of the diode's two). In the diode, electrons were strongly attracted to the positively charged plate; but a small negative charge on the triode's grid could interfere with that by repelling emitted electrons back toward the filament before they got very far. Small variations in the electrical potential applied to the grid could have large effects on the current reaching the plate—but the way the plate current varied closely followed the variation of the grid voltage.

That fact was particularly useful in radio and sound reproduction because it provided the basis of an amplifier. A weak electrical signal could be applied to the grid, and a larger but otherwise very similar one obtained at the plate. By applying that signal, the output of one vacuum tube, to the grid of another, it was possible to make multistage amplifiers that could produce really big outputs. This was suitable, for instance, for filling a large room—from a small input such as that from a phonograph cartridge. Much research over the ensuing decades was devoted to designing amplifiers—both the tubes themselves and the associated cir-cuitry—that could produce a large gain (amplification factor) with little distortion, so that reproduced and amplified sound was as similar as pos-sible to the original.

But that wasn't the only thing triodes could do. In different kinds of circuits, they could act not as amplifiers but as switches, giving a certain fixed output if the input was above a certain level and none if it was below that level. In other words, it could be used to make a *digital* cir-cuit, producing a yes-no output. (A conventional amplifier is an *analog* circuit, producing an output as much like the input as possible.) Such a digital vacuum tube circuit could act much like the relays I talked about earlier, but much faster because electrons are so much less massive than everyday mechanical objects.

You may now be thinking, "If a good amplifier can produce an output closely analogous to the input, why not make an *analog* computer? It would code real-world variables as voltages that vary in the same way and then combine them in appropriate ways to get an output voltage that varies like the solution to a problem." The short answer is that you can, and some have. You may well have heard of analog computers, but you've probably also noticed that almost all the computers (and a lot of other things) now surrounding us are digital. Digital handling of information has some very significant advantages. The world is full of stray signal sources; what radio listeners call static is caused by things like lightning and vacuum cleaners running nearby. Those get added onto the desired signal and distort it. But if the signal is digitized, by expressing each piece of it as a number, such extraneous noises have no effect unless they're big enough to "trip a switch" and change one number to another.

For this and other reasons, computer developers have tended to gravitate toward digital methods, even back in the 1930s and 1940s. The Harvard Mark I, as already noted, was a digital machine using electromechanical relays for switching. Around the same time, others, such as the American mathematician and physicist John V. Atanasoff, were making their own computers using vacuum tubes for the switching elements. Other first-generation computers using vacuum tubes included ENIAC (Electronic Numerical Integrator and Calculator, developed by J. Presper Eckert and John W. Mauchly at the University of Pennsylvania), and UNIVAC (Universal Automatic Computer, built by the same team for the US Bureau of the Census in 1951). Those demonstrated dramatically the advantage of vacuum-tube computers: they were a thousand or more times as fast as their electromechanical counterparts.

But that was by no means the end of the line. By the time Eckert and Mauchly were unveiling UNIVAC, the second of those "major tributaries" had already started carving out its own channel.

On the face of it, solid-state physics might seem about as far from electronics (as I've defined it so far) as you could get. Whereas solid-state physics deals with matter in its densest and most rigidly ordered form, electronics requires free electrons in spaces containing as close to nothing as possible. Yet the two fields turned out to offer quite different ways of doing the same things.

Ironically, the very first radios used a crude sort of solid-state rectifier. Certain kinds of crystals, such as the lead-containing mineral galena, allowed current to flow much better in one direction than the other, and were the basis of early radios, which were consequently called *crystal sets.* But they didn't work particularly well, and were a pain to use. I can testify to that from personal experience, having used one that belonged to my great-uncle. Making it work required painstakingly moving a tiny flexible wire called a *cat's whisker* around on the surface of the crystal, looking for a spot that produced the desired result. Vacuum tube rectifiers were so much easier and more reliable to use that they quickly replaced natural crystals as rectifiers in radios.

In the 1940s the English-American physicist William Shockley and American physicists John Bardeen and Walter Brattain at Bell Telephone Laboratories (an industrial establishment remarkable for its encouragement of basic research) were experimenting with the properties of semiconductors. As the name suggests, semiconductors are a class of solid substances that conduct electricity better than insulators (like glass) but not as well as the highly conductive metals (like copper).

Semiconductors include elements like germanium and silicon. Shockley's group found that their semiconducting properties could be greatly enhanced by "doping" them with small amounts of certain impurities. The basic reason that a semiconductor acts as such is that, like a metal, it contains free charges that can move—that is, carry a current—when an electric field is applied; but it doesn't have very many, so it doesn't carry a current as strongly as a metal. Adding a tiny amount of antimony to a germanium crystal causes it to have more electrons that are loosely bound and are therefore attracted toward the positive terminal when a voltage is applied to the crystal. This enables the crystal to conduct more current. Such a semiconductor, doped to have an excess of free electrons, is called an *n*-type semiconductor (because the extra electrons have a negative charge).

Other additives can create another type of more-conductive semiconductor called *p*-type because its excess charge carriers are (at least effectively) positively charged. I've always found this type a little harder to get a mental handle on, but the essence of it is this: if you add a little iridium to a germanium crystal, it tends to lock up some of the loose electrons already in the crystal, creating "holes" that can be filled by any introduced electrons that pass nearby. When a voltage is applied to a *p*-type semiconductor, the conductivity is again increased; but in this case it's convenient to think of the holes as being the charge carriers, and since they're positive, they drift toward the negative terminal.

If you put a piece of *p*-doped germanium against a piece of *n*-doped germanium, current will flow across the junction quite well in one direction and hardly at all in the other. Such a germanium diode turns out to be a much better rectifier than the old-fashioned galena diode, and also much easier to use because you don't have to fuss with trying to find the "sensitive spot" with a cat's whisker and trying to get it to stay there. You simply fasten one wire to the *p* side of the junction and another to the *n* side and you have a highly reliable rectifier that you can connect positively into its circuit by simply clipping or soldering external wires to the wires (or leads) that are permanently attached to the diode. I can vouch for this, too; after fussing with the cat's whisker on my great-uncle Bill's old crystal set, I simply used alligator clips to connect a germanium diode—physically, a tiny glass capsule with a *p-n* junction inside and a wire coming out of each end—in place of the old crystal. Instantly the set's operation became completely reliable, with no fussing. (The *quality*, of course, still left much to be desired, because this type of radio is intrinsically pretty primitive.)

The great contribution of Shockley and his colleagues (for which they shared the Nobel Prize in physics in 1956) was their discovery of what happened if they combined what amounted to two *p-n* junctions in opposite directions in a single package—for example, a piece of *p*-type germanium sandwiched between two *n*-type pieces (or vice versa), with wires attached to all three elements. Such a device, called a *transistor* (transferring current across a resistor), turns out to be closely analogous to a triode vacuum tube.

One of the two outer pieces of the sandwich is called the *emitter* and the other the *collector*. The things emitted or collected are electrons or holes, depending on whether those pieces are *p* or *n* type. The piece in the middle, with the opposite type of doping, is called the *base*. The emitter, base, and collector correspond respectively to the filament, grid, and plate of a triode. Just as a small change in the voltage applied to the grid of a triode can produce a correspondingly large change in the current from filament to plate, so can a small change in the voltage applied to the base of a transistor produce a correspondingly large change in the current from emitter to collector.

But there are several important practical differences between vacuum tubes and transistors. Part of a vacuum tube has to be red-hot to "boil off" electrons, which means you have to wait for it to "warm up" before it can do anything useful, and you also have to provide some means of disposing of large amounts of waste heat all the time it's running. The transistor starts working as soon as you turn it on, and can work with quite small currents (except where power is specifically the goal, as in the final stage of an audio amplifier), so it runs relatively cool. The most obvious advantage of transistors, though, is that they are so much smaller than equivalent vacuum tubes, which allows devices using them to be much smaller.

The transistor was, at least nominally, invented in 1948, but it took a few years to learn to make them well enough and reliably enough for general use. By the mid-1950s transistor radios had become common, with great novelty value because they were often smaller than most books, could be used practically anywhere, and came on immediately with low-voltage batteries. But the ones my grade-school classmates had didn't sound very good, and it took me awhile to overcome a bias against them as cheap substitutes for tubes as they eventually found their way into better and better circuits.

And for computers the importance of miniaturization was even more important, as machines that used to fill huge rooms could now fit into boxes the size of refrigerators—or even smaller. And here we see the input of yet another tributary: space exploration. When the Soviet Union launched the satellite Sputnik I in 1957, in the midst of the chronic hysteria called the Cold War, they also launched a "space race" in which they and the United States struggled to outdo each other with feats of rocketry and telemetry. Since sending things into orbit or beyond with rockets is extremely expensive, and onboard instrumentation required sophisticated circuitry, the space program provided a powerful incentive for research toward making smaller and smaller electronics.

Shockley, Bardeen, and Brattain, when they first conceived the transistor, probably would have been amazed if they could have seen how far that trend toward miniaturization would go in just a few short years—and how it would get there. By the end of the 1960s, solid-state physicists and engineers had learned to make "integrated circuits," tiny silicon chips (think fingernail-sized) incorporating dozens or hundreds of transistors and associated components such as resistors and capacitors. All those parts were "wired" together in a single unit whose components couldn't be separated—or even seen, without a microscope.

Work on ways to make such devices even more minuscule continued in many places. One day in the 1980s I visited Bell Labs and was shown the work of several different groups, all working on different methods (obviously all quite different from traditional manufacturing methods!) for making even more microscopic integrated circuits. By the late eighties, Very Large Scale Integration (VLSI) could cram hundreds of thousands of components onto a microchip much smaller than your little fingernail. (See plate 4.)

A side benefit of the components being so small and close together is that signals don't have to travel very far to carry out their operations, and so those operations can be carried out with speed that would have been considered the wildest sort of dream a couple of decades earlier. As I write this, the power and speed of computers continues to increase so rapidly that the latest and greatest thing is commonly regarded as obsolete within a year or two. Any commuter train car now contains numerous passengers casually working on computers far more powerful than the one that served an entire university when I was in graduate school, yet smaller—sometimes *much* smaller—than a small notebook.

And microchips (or microprocessors) have found their way into a vast multitude of everyday applications that might seem to have little to do with computing: automobiles, microwave ovens, telephones, television sets, stereo systems, CD and DVD players, MP3 players, cameras—even sophisticated sneakers and prosthetic knees. Almost anything even partly electrical is likely to contain one or more microprocessors as part of a control system that allows it to carry out a great variety of operations according to the user's wishes, often programmed in advance.

Perhaps the most important category of those applications involves one last convergence. Way back in 1831, the American physicist Joseph Henry was inventing the telegraph. Samuel Finley Breese Morse usually gets the credit, and in 1840 he got the patent; but the true story is a good deal more complicated.[6]

Joseph Henry was born into a poor family in Albany, New York, in 1797. With little schooling and the need to go to work quite young (he was apprenticed to a watchmaker at the age of thirteen), he was largely self-educated but still wound up teaching at Albany Academy before he was thirty. He was intrigued by reports of the Danish physicist Hans Christian Oersted's discovery of the interaction between electricity and magnetism (made by putting a compass near a wire carrying a current). Henry then did his own experiments with all shapes and sizes of electromagnets.

In particular, he discovered that if he put a small electromagnet at one end of a long pair of wires, and a battery and switch at the other, closing the switch could activate the electromagnet and attract an iron bar attached nearby. A spring suitably attached to the bar could pull it away when the current through the electromagnet was turned off. If the current was turned on and off in a particular rhythm, the bar at the other end of the line would move against and away from the magnet with the same rhythm. That's a telegraph: a coded message could be sent by tapping out a rhythm on the switch (or "key") at the battery end, and someone who knew the code could read the message by watching the motions of the bar against the electromagnet at the other end.

It wasn't good for very long distances because the resistance of the wire made the current, and consequently the action of the electromagnet, weaker as the wires got longer. But it didn't take Henry long (about four years) to find a way around that. If he attached another switch to the iron bar at the receiving end, he could make the wires long enough so the current could just barely actuate the receiver. When that happened, it closed a new circuit with a battery that could transmit a fresh but identical signal down another pair of wires to another receiver—and so on, as far as anyone liked. You probably recognize from that description that Henry had just invented the relay, and in fact this application probably shows where the name came from. The action of those relays is much like that of runners in a relay race, carrying a baton a certain distance and then passing it to a new runner to carry on.

With a sender at one end, a chain of relays, and a receiver to deliver a message at the other end, Henry had quite clearly invented a telegraph system of considerable power. But he was too altruistic for his own good and never patented his inventions.

Meanwhile, Morse, a privileged Bostonian who graduated from Yale, studied art in England and lived there awhile as a moderately successful artist. He became interested in electrical experiments through conversation with a fellow passenger named Jackson on an ocean voyage. He wanted to build a telegraph but had little idea how to go about it until he met Henry and pumped him for information. Despite his lack of scientific sophistication, Morse's stubbornness enabled him to drum up support for the work needed to make the telegraph a practical tool. He got a patent for it and somehow persuaded Congress to finance a telegraph line between Baltimore and Washington, DC, which in 1844 carried its first message, "What hath God wrought?" He never publicly acknowledged what Henry had wrought, and while we are told that Henry didn't mind not sharing in the financial rewards, he did provide testimony in a legal dispute between Morse and Jackson that clearly proved he deserved (and wanted) credit he wasn't receiving.

These days the telegraph is less familiar to most people than its more versatile spin-off, the telephone. The telegraph basically is a device for telling a switch at the other end of a cable to turn *on* or *off*. But what if, instead of a simple on-off switch, you had something that varied in a much more complicated way, and could be used to make something at the other end of the cable vary in the same way?

Alexander Graham Bell got interested in such problems in a roundabout way.[7] He was born in 1847 to a Scottish family with a long history of interest in human speech. His grandfather and father studied sound, and in the late 1860s Alexander worked closely with his father in the education of deaf children. His family, including Alexander, also suffered health problems in Scotland. In 1870, after he lost two brothers to tuberculosis, they all moved to Canada (which seemed to help). He later continued his research as a professor at Boston University and got the idea of modifying the telegraph to send *sound* from one place to another at the speed of light.

The trick was to convert sound waves to an electrical signal by making them vibrate a membrane whose motions would induce an electrical current that varied in the same way. A device for doing that is called a microphone, or, in telephone parlance, a transmitter. At the other end was a receiver (which we might also call, depending on its size and power, an earphone or a speaker) that did the same thing in reverse. The varying electrical current from the transmitter caused a membrane or diaphragm to wiggle, pushing on air and producing sound waves similar to the original. (The same device can function as either a microphone or a speaker, since either one involves a vibrating membrane with an electrical current on one side and sound waves on the other. Useful devices are designed to be much more efficient at one kind of conversion than the other, and that is what determines whether we call a particular gadget a speaker or a microphone.)

With support from the ever-helpful Henry, Bell made the idea a reality, and in 1876 (some twenty years before Marconi's radio work), he patented the telephone. (The famous first transmission of "Watson, please come here. I want you" apparently came about by accident. As Isaac Asimov tells the tale, Bell was working on a telephone prototype and burst out with the famous request when he spilled battery acid on his pants while the device was turned on.)[8]

Use of the telephone caught on and spread rapidly. A great deal of research over subsequent decades was devoted to refining and expanding its capabilities: making it more reliable, usable over long distances, easier to use, and possible to use in new ways such as conference calls. More recent developments include the use of synthesized voices, voice recognition and interpretation, and touch-tone codes to automate many routine operations and eliminate the need for a human employee at the business end of many calls (though as almost anyone can testify from experience, that isn't always an improvement!). And, of course, the use of telephone lines to send digital data followed.

In this we have, in a sense, come almost full circle. The original telegraph was a purely and simply digital device, but quite slow by modern standards. Today data are commonly expressed and manipulated in a digital code made of switches set to ON or OFF. But the switches are now electronic, and enormously faster than the ones used by Morse and his contemporaries. Add the capability to transmit such data by telephone lines, and the ability to use such lines to connect many centers of data production and processing, and you get, in a remarkably short time, the modern internet—something that exceeds by many orders of magnitude anything that Jacquard or Babbage or Henry or Morse or Bell could have imagined.

In fact, almost the entire science fiction community pretty thoroughly missed the boat on that one. I'm hard put to think of any science fiction story from before 1970 or so that imagined that computers would ever be anywhere near as small, powerful, ubiquitous, and interconnected as they've already become. In 1939, a fact article by Leo Vernon in *Astounding Science Fiction* (a magazine often credited with revolutionizing modern science fiction) described a "dream machine" of a computer that might be built sometime in the future: "[It] may fill an entire building. It will be operated from a central control room made up entirely of switchboard panels, operated by trained mathematicians, and an automatic printer giving back the results. . . . [A scientist's long tables of numbers] are typed by a stenographer on a special machine which punches them on a tape. The tapes and the directions giving the order of the use of the numbers are turned over to the operator. The tapes are fed into a slot, a switch pulled, and in a few minutes every number has been multiplied by all the other necessary numbers. . . ."9

This was in a magazine whose writers were widely acknowledged to be some of the most imaginative, far-seeing in the world. Try to imagine how surprised Vernon and his editor, John W. Campbell, would have been had they known that a mere fifty or sixty years later, at least one computer far more powerful than his "dream machine" would sit on the desk of practically any scientist in the country. And that a large percentage of private homes would also have at least one and often several, routinely used for tasks ranging from bookkeeping to everyday correspondence to grade school homework and games. Imagine further how astounded they would have been to learn that most of those computers would be connected together in ways allowing practically instant communication and access to what amounts to the world's largest (though quite imperfectly edited) encyclopedia.

One thing that would *not* have surprised them, given those facts, was that all those technological changes are revolutionizing practically every aspect of the way people live, learn, shop, and work. Major technological changes always do. None of this particular cluster of them would have come about without the convergence of all those seemingly distinct lines of development that I have discussed in this chapter. And the process is far from over. We have not arrived at the metaphorical ocean; we are instead immersed in a mighty current that has developed from the confluence of several smaller ones and is still rushing onward, gathering force—and new tributaries—as it goes. Where it will eventually lead we can only imagine; but we must try to imagine as well and as soon as we can, to be prepared for what lies ahead.

No analogy is perfect, of course, and this one is no exception. I certainly would not want to suggest, for example, that electronics is only important for contributing to the development of computers. It exercised a large influence on many other fields as well, from communications to entertainment to navigation to medicine. The actual relationships among different lines of work in science and technology are perhaps more like the threads in an elaborate tapestry. But if you tried to look at them all at once, the pattern would be too overwhelming to comprehend. Sometimes it's more manageable, and therefore more helpful, to concentrate on a few threads at a time. In doing that, it's helpful to revert to the converging-streams model, with the electronics stream in this chapter, for example, representing not the whole of electronics, but rather the part of electronics that fed into and influenced the development of computers.

The computer stream, as we shall see, promises to converge with practically everything else, and some chapters hence we shall look at some of the promises and perils that will shape our overall future as a result of all those convergences. That is one of the central concerns of this or any other age. But before we can do that, we must look at some more major streams.

CHAPTER 2
AVIATION AND BIG BUILDINGS

E veryone has heard of Orville and Wilbur Wright, and if you haven't actually looked at the history, it may seem that the whole current of aviation began abruptly at Kitty Hawk like a stream gushing from a spring. In reality, their work, too, had origins still further back—in fact, it too represented a convergence of several earlier currents, some of which resulted from still earlier convergences.

The Wright brothers' prime accomplishment—and it is clearly theirs, and very important—was the first controlled, self-powered, sustainable flight by a heavier-than-air machine that carried a person. Others had made powered vehicles before them; others had made long flights in lighter-than-air devices (balloons); others had soared in gliders or sent bombs rocketing to their targets. The Wrights were the first to put all the pieces together, gaining the ability to make heavier-than-air craft fly by their own power rather than the whims of the winds, and go where their pilots wanted them to go.

But how did they reach that point?

Attempts to achieve powered flight go way, way back. An ancient Greek myth tells of Daedalus, whom we might call an engineer, making birdlike wings of feathers and wax to enable himself and his son Icarus to escape from a prison where all land and sea routes were blocked. Things didn't go as well as he'd hoped; Daedalus got out all right, but Icarus flew too close to the sun and the wax holding his wings together melted. He plunged to his death in the sea, and the remorseful Daedalus hung up his wings, literally and figuratively.

We can't know how much, if any, grain of truth is buried in the myth, but we do know that people have long watched with envy the flight of birds. Quite a few have tried to imitate it—with results more like those of Icarus than Daedalus. Without an understanding of the fact that air is a substance with such properties as mass, temperature, and pressure (not proved until the invention of the mercury barometer in 1643), trying to build birdlike wings to be strapped to arms, or incorporated into machines called *ornithopters*, was pure guesswork. Many tried, but either they never got off the ground or they returned to it immediately (and often tragically). And though such far-seeing thinkers as Roger Bacon (1214–1294) and Leonardo da Vinci (1452–1519) vaguely saw the possibility of human flight and tried to imagine how it might be done, they lacked the means to figure out just what it would take. Had they been able to do so, they might have saved many years of blind alleys (and some lives) by demonstrating theoretically that birdlike flight just isn't practical for humans. Given Earth's gravity and atmospheric conditions, flying like birds would require prohibitively large wings and far more muscle power than any of us could supply. And the kind of motion those wings would have to make is so complicated that a machine that could do it would be extremely difficult to build.

Ironically, as Alan Howard Stratford observed in the *Encyclopaedia Britannica*, our ancestors' long preoccupation with trying to imitate the most obvious model may well have long delayed the development of flight as achieved by the Wrights and their successors.[1] Birds could do it all, but that was too much for a first step. What ultimately led to the Wright brothers' accomplishment was breaking the problem into two smaller ones. Prior to their success, no prescient project manager said, "You work on this one and I'll work on that one, and then we'll put them together and have manned heavier-than-air flight." Instead, different people worked on the two things that were needed for reasons of their own—and then those currents could converge, bringing the solutions to two smaller problems together to solve a bigger one.

ON WINGS OF . . .

While all birds fly at least some of the time by flapping their wings, some of them do much of their flight *without* flapping—and that was a crucial observation. Before people could succeed at *powered* flight, they needed to understand on at least some level what enables vultures to circle for hours with little if any flapping, and albatrosses to spend practically all their time far above the oceans. In other words, they needed to understand flight in its simplest form before trying to add power.

Not everyone understood this, and through the nineteenth century some experimenters persisted in trying to achieve powered flight by adding steam engines—already developed for railroads, but much too heavy for flight—to flimsy and poorly understood structures. One of those, a steam-powered plane designed and built by Aleksandr Mozhaysky in 1884, was claimed by Russia as the world's first airplane flight on the strength of a launch from a ski ramp and a few seconds in the air before crashing. A French engineer, Clément Ader, in 1890 also got off the ground for a few dozen yards and lived to tell the tale; but it was, as Stratford put it, more a "hop" than a controlled flight.[2]

Learning control, especially for the critical times of takeoff and landing, required learning, through both theory and practice, what makes vultures and albatrosses "tick." George Cayley, an Englishman some-times called the father of aeronautics, dealt in both, building glider models of various shapes and sizes and analyzing airflow through the long period from 1792 to 1857.

By then it was well understood that air was a gas, and scientists knew a good deal about how gases behave. One of the most important facts was that when a gas flows around a kind of structure called an *airfoil*, with the top surface convex and the bottom nearly flat, an upward force (called *lift*) is produced. If air is blown toward the front of an airfoil (such as an

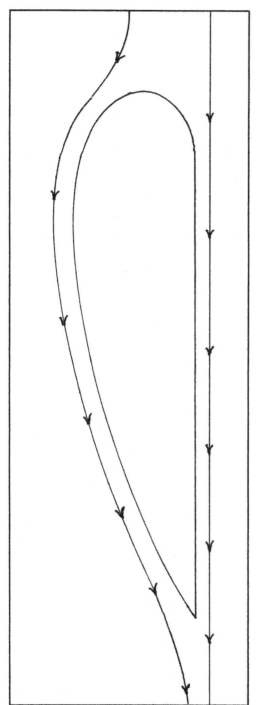

Figure 2. Flow of air around an airfoil, such as an airplane wing. Whether the wing is moving to the right or the air is flowing to the left, relative motion over the top is farther and therefore faster than along the bottom, resulting in an upward force called "lift." (Diagram by the author.)

airplane wing), the greater curvature forces air flowing over the top sur- face to travel farther than that along the bottom surface. Thus it has to travel faster. When a stream of air travels faster, its sideways pressure is reduced, so pressure above the wing is lower than that below and there is a net upward force. The strength of that force depends on the exact shape of the wing.[3]

The important concept here is the *relative* motion of the wing and the air. It doesn't matter whether the wing is held stationary and air is blown over it, as in a wind tunnel, or the wing has somehow been placed into motion through the air, as when it is attached to a soaring eagle or a sailplane. Thus the effectiveness of a new shape could be tested in a wind tunnel before anybody tried to fly with it.

Early experimenters with gliders didn't have wind tunnels, but tested different designs with small models and sometimes by building and attempting to pilot full-scale versions. The most influential of those intrepid souls was the German engineer Otto Lilienthal, who built a series of gliders, refining the art enough to make gliding a sport of some popu- larity during the 1890s.

The big problem with early gliders was not getting airborne (that could be achieved by running off a cliff, with which perhaps the main dif- ficulty was psychological), but controlling what happened afterward. Lilienthal's original idea was to control his gliders' flight by shifting his body weight, but it didn't work very well. Toward the end, he began trying to utilize his understanding that the exact shape of an airfoil affects its flight characteristics. For example, if the left and right wings of a plane had different shapes, the one producing the greater lift would tend to rise, causing the plane to roll into an attitude with one wing higher than the other—and its lift acting at an angle to the vertical, causing the plane to turn. For this effect to be useful in steering, of course, you would have to be able to change the shape of a wing in flight—and that's exactly what is done in modern planes, where movable flaps and ailerons change the effective shape of a wing to make the plane behave in certain ways.[4] Other adjustable airfoils, both horizontal (*elevators*) and vertical (*rud- ders*), add further control of steering, climbing, and descent.

Lilienthal died in a crash while testing such a configuration in 1896, but his work was carried forward by others who had been following it, such as Octave Chanute, Samuel Pierpont Langley, and Orville and Wilbur Wright. Langley, who among other things was an astronomer and the third secretary of the Smithsonian Institution (despite having never gone to college), did extensive experiments with wing designs mounted on the end of a whirling arm and made significant progress in formulating aerodynamic principles.

He was also interested in powered flight and tested a number of models powered by stretched or twisted rubber bands and later by steam or gasoline engines. The Spanish-American War got the government interested enough in the possibility of military aircraft to give him a sizable grant for research. That work included model tests that encouraged him to build full-scale machines, but those didn't fare so well. After two dramatic and discouraging failures, Langley lost both governmental and popular support, was subjected to widespread ridicule, and got no more chances. An editorial in the *New York Times* mocked his waste of public money and predicted that if man ever flew, it would not be for a long, long time.[5]

A few days later, Orville Wright did it.

DRIVING FORCES

Obtaining a suitable engine was by then the biggest hurdle, and the solution came in large measure not from people trying to fly, but from others who had for some time been after quite different goals. Technically, the earliest steam engines were simple novelties built by Hero of Alexandria in the first century CE, but it was many centuries before anybody began putting steam to practical use. Even the earliest practical engines, used by Thomas Savery in 1698 to pump water out of mines, were considerably more complicated than Hero's aeolipile. Steam engines are "external combustion" engines, in which the working fluid that pushes a cylinder is vaporized by a heat source outside the expansion chamber. Thomas Newcomen, an English blacksmith, improved Savery's design, and his version was used for water pumping for half a century.

James Watt, a Scottish engineer, too often thought of as inventing the steam engine from scratch, was home-schooled because of poor health, subject to migraines, and suspected of mental retardation. After an increasingly rough childhood, he eventually wound up in London as an instrument-maker's apprentice. He didn't stay at that long enough to meet the city's licensing requirements when he returned home to Glasgow (where a statue of him now stands in the city's largest public square), but the city didn't run the University of Glasgow and he got a job there.

While at the university, in 1764, he was asked to repair a troublesome Newcomen steam engine, and that got him thinking about its inherent inefficiencies. He came up with one major and several secondary improvements, and, with a partner, went into business making and selling steam engines. In less than twenty years the Watt engine had completely replaced the Newcomen version and was coming into such widespread use that it became a major cause of the industrial revolution.

Here, too, another tributary played a significant role. At this time England's most important industry, textiles, was being mechanized. Manufacturing machines required power to run them, and up to then, that meant they had to be located where fast-flowing streams could turn waterwheels. The Watt engine eliminated that requirement, making it practical to build factories nearly anywhere by having steam engines on the premises. Furthermore, those factories and the machines they housed could be *big*, allowing the concentration of more manufacturing activity in one place than ever before.

The result was thus: Watt made better steam engines; the mechanization of textile making created a large and growing market for them; factories largely replaced home crafts by individual artisans; and workers concentrated in rapidly growing cities, complete with slums. It was a dramatic illustration of how a seemingly simple convergence of technologies can radically reshape an entire way of life.

Those early engines were all stationary, permanently housed in buildings, with systems of belts and pulleys carrying the power they produced to equally stationary machines. They were also big and heavy, since they required several separate components such as (typically coal-burning) furnaces, boilers, cylinders, pistons, condensers, and sometimes superheaters,[6] all made of metal (typically cast iron). This wasn't much of a problem when they just sat in one place, but it imposed serious limits on how they could be used for transportation.

Thus the earliest applications of steam power to vehicles were to *big* vehicles: trains and ships (or at least large boats). A Jesuit missionary claimed to have seen a steam-powered land vehicle in China in 1681, but documentation is sparse. A French artillery officer named Nicolas-Joseph Cugnot built a huge, three-wheeled munitions carrier in 1769; from our point of view it's hard to say whether it should be called an automobile or a locomotive, though we might lean toward the former since it ran without tracks. Tracks were adopted early as a means of controlling those massive machines (and absolutely necessary when several cars were joined into a train). But the first tracks were not up to the job and led to the premature failure of such pioneering efforts as those of the Cornish engineer Richard Trevithick.

Trevithick demonstrated working trains as early as 1801, but he had more good ideas than he could do justice to and never fully succeeded with any of them. However, others were exploring similar territory, and by 1825 the English inventor George Stephenson had built and opened the first public railroad to haul passengers and freight. Soon railroads proliferated all over Europe, making long-distance overland travel unprecedentedly practical.

Meanwhile, still others were applying steam power to travel on water. Here again, the popular wisdom is wrong: Robert Fulton did not invent the steamship; he was merely the first to make a commercial success of it, so he's the one a casual reader of history remembers. A decade before him, John Fitch, an American inventor born in Connecticut and plagued by misfortunes throughout his life, hatched the idea in 1785 and by 1790 he had a steamboat running a regular ferry service between Philadelphia, Pennsylvania, and Trenton, New Jersey. It didn't attract enough business to thrive, and the business folded within a couple of years when its one boat was wrecked in a storm.[7]

So priority of invention clearly goes to Fitch, but that does not minimize the significance of Fulton's contribution. History is full of "ideas before their time," discoveries or inventions that were discovered at least once, failed to take off, and thrived only when they were resurrected in more favorable times or by cleverer and/or luckier entrepreneurs. Only the ones that thrived are remembered by most of us.

Fulton, born in Pennsylvania, worked as a jeweler's apprentice and later as an artist (among other things, he painted a portrait of Benjamin Franklin), but then became interested in civil engineering. He studied in England, then stayed on to work on canal navigation there and, later, in France. That led to interest in powering boats with steam-driven paddle wheels. He didn't get far with the idea in France, but after returning to America he built the *Clermont* in 1807, a boat good enough to make regular, profitable runs between New York City and Albany. The *Clermont* was soon joined by a fleet of companions, and for the first time, the door was wide open to boat commerce not dependent on oars or capricious winds.

Steamboats still provided little incentive to build engines that could meet the additional requirement imposed by flight: high power combined with light weight. But by the end of the nineteenth century, several important currents were converging that led to engines *almost* good enough.

BRINGING IT ALL TOGETHER

Remember Cugnot's big steam tricycle, mentioned earlier? It was the progenitor of railroads, which were its descendants of most lasting import. However, another line also sprang from that source (something we might view as a technological *divergence*). Trains on rails soon took on great economic importance. But even before that happened, some people were experimenting with single vehicles that could run without rails. Their steam engines were big, so the vehicles were big, and today largely forgotten. Paris had steam buses (or coaches) around 1800, and various inventors and entrepreneurs built and operated variations on the theme in several countries beginning around 1790.

Steam coaches had their heyday, such as it was, in the 1830s, but that was when railroads were really catching on and proliferating. Coaches had trouble competing, and the climate of public opinion was against them. People feared boiler explosions and disliked the vehicles' noise, smoke, and impact on road surfaces. Furthermore, the entrenched horse-carriage lobby did everything it could to interfere with horseless carriages, such as pushing for legislation subjecting them to ridiculous speed limits and exorbitant tolls, particularly in Britain.

Later in the nineteenth century, despite continued popular opposition, steam cars enjoyed some renewed enthusiasm, including some small enough for just two people and capable of speeds as fast as 20 mph. Such things were made possible by advances in two other fields: metallurgy and toolmaking. The ability to make strong steels (which could withstand high pressures) and light aluminum structures, coupled with an unprecedented ability to make machine parts to high precision, made it possible to produce relatively light steam engines—though still not light enough for aircraft.

But the new materials and techniques could also be used to make a new, fundamentally different kind of engine: the internal combustion engine, fueled by gasoline. Instead of using external combustion to heat a confined but chemically inactive fluid to drive a piston, as in a steam engine, the internal combustion engine moved the heat source *inside* the driving cylinder. It mixed a gaseous or vaporized fuel (such as gasoline) with air, exploded the mixture with an electrical spark, and used the explosion products themselves as the expanding gas to drive the piston.

The first such engine that worked was built and applied to a horseless carriage in 1860 by the self-educated Belgian-French inventor Jean-Joseph Étienne Lenoir. Sixteen years later, Nikolaus August Otto, a German traveling salesman, built a much more efficient version. This used the motion of the piston to operate valves drawing fuel and air into the cylinder and compressing them, igniting them with a spark, and then forcing the exhaust gases out to start the process over.

This Otto cycle (so-called even though the French engineer Alphonse Beau de Rochas had described the principle and obtained patents several years earlier) became the basis of many subsequent engines, especially in applications where small size and light weight mattered—as in private automobiles. Making it work depended on the materials and methods already mentioned, coupled with sufficient control of electricity to produce precisely timed ignition sparks. Getting rid of external furnaces and condensers allowed gasoline engines to be even smaller and lighter than steam engines of comparable power. Explosive ignition by spark made them much more responsive to quick control. All of those properties quickly made them the engine of choice for automobiles, but before they could be manufactured in large numbers, they needed plenty of fuel in gaseous or volatile liquid form.

That need was fulfilled by yet another converging current: in the late nineteenth century, large deposits of oil were discovered. Chemists learned to refine it into just the sorts of fuels that were needed for the new engines. With fuel readily available, lots of inventors and businessmen got into the act, experimenting with new car and engine designs and competing for economic advantage.

That's where things stood at the turn of the century, when Orville and Wilbur Wright were at the point of needing an engine just a bit better than any then available. The brothers, who were brought up in a minister's prim and proper household, with little formal education (neither finished high school), had lots of ingenuity, patience, and a methodical approach to solving problems. They also shared an active interest in bicycling and gliding. They put that interest to practical use by making a living as bicycle repairmen, and in their spare time worked in an unprecedentedly orderly fashion toward powered flight.

Unlike most of their predecessors, they did not make the mistake of trying to leap directly to that goal. Instead, they systematically studied the problems of designing workable wings and aerodynamic control surfaces. They winnowed through the vast chaos of earlier experimental results, trying to sift the relatively few useful facts from the surrounding multitude of inaccuracies and errors. Even more important, they did a great many experiments of their own, initially mounting test surfaces and measuring instruments on bicycles, then building a wind tunnel, and finally testing kites and gliders, first unmanned, then manned. Their hometown, Dayton, Ohio, was not well suited to those field tests. So they set up camp at Kitty Hawk, North Carolina, which had reliable winds and soft sand for hard landings. Despite a couple of near-catastrophes, by 1902 they had the world's first truly practical glider, capable of staying aloft for prolonged periods and actively controlled by its pilot.

And only then were they ready to tackle powered flight.

They found no engine on the market that quite met their needs, and no manufacturer willing to commit to developing one. So they, and a capable assistant named Charles Taylor, built their own. On the morning of December 17, 1903, they made three successful flights (the longest lasting 59 seconds and covering 852 feet against a 21 mph headwind) before a wind gust damaged their plane. The press and public reacted with indifference, and the brothers made no attempt to change that. They were too busy continuing their research and further refining both their machines and their flying skills. Now feeling less need for soft dunes to land on, they rented a ninety-acre cow pasture closer to home as a test field.

Eventually others started to take notice. The first published eyewitness account of manned flight appeared, oddly enough, in the January 1905 issue of a beekeeping journal published in Medina, Ohio,[8] and soon there was evidence of potential competitors beginning to emerge. The Wright brothers finally began thinking about legal protection and financial return, applying for patents and removing their new work from public view. They persuaded a reluctant War Department to let them try to develop a military plane to seemingly impossible specifications, and by the fall of 1909 the army had acquired its first plane and Wilbur had established its first flying school.

And growing numbers of other people got into the act. Aviation lent itself to showmanship, especially at a time when it still had much novelty value and had not yet demonstrated much practical usefulness. Much of the flying done between 1910 and 1914 took the form of exhibitions and competitions—what some have called the *aviation carnival.*

World War I changed all that. Suddenly governments, engaged in combat, became very interested in developing the military potential of the new technology. They put planes to work in roles ranging from combat to reconnaissance to message and supply delivery, and they funded research to develop new kinds of planes and optimize them for specific tasks. Aviation and aeronautical research became big business. As often happens, war stimulated a burst of new development, and for four years, practically all of the burgeoning activity in aviation was aimed at winning the war.

When the war ended, so did that activity. Thousands of pilots and air-planes were suddenly unemployed. Some of the planes were simply abandoned or destroyed; others were bought cheaply as surplus by pilots who became barnstormers, entertaining the public, keeping popular interest in flight alive, and creating the beginnings of interest in civilian air transport. In 1918 the US Post Office started flying some of the mail, and gradually a new market for aircraft developed, along with new research to make planes safer, more reliable, and more comfortable. In the late 1920s the Post Office began contracting some of its mail delivery out to private aviation companies, and some of them began also carrying passengers.

Initially commercial passenger service was expensive, not very safe, and not heavily used. But as safety, reliability, and comfort continued to improve, it became more and more popular and competitive with other forms of transport. Long-distance flights, even across continents and oceans, became possible and then routine. New navigation systems, using radar and other forms of electronics, made it possible to take off, fly, and land safely in almost any weather. Jet engines, a fundamentally new type inspired by new understanding of fluid dynamics that allowed much higher speeds, were first applied to aircraft in the late 1930s. They soon became standard for fighters, and in the 1950s began to appear in commercial aircraft. Manufacturers learned to build much bigger airframes and engines to drive them.

And by 1970 or so, flight had become the safest and most popular form of medium-to-long-distance transportation, with planes routinely carrying up to several hundred passengers anywhere on the planet in no more than a few hours.

MEANWHILE, BACK ON THE GROUND . . .

The urge to build big structures also goes way back. The Bible tells of the Tower of Babel, an ambitious project that made God fear that too much cooperation was letting humans get "too big for their britches"; so he divided them into mutually incomprehensible language groups to make that harder. Here again it's hard to guess how much literal truth lay behind the legend—many scholars now view the story as an early attempt to explain the diversity of languages—but it at least demonstrates the desire for huge structures.

We're quite sure that several dynasties of Egyptians built pyramids hundreds of feet high, as did the Maya, Aztecs, and Incas a good deal later. Those were essentially big, carefully organized piles of rock for monumental or ceremonial purposes; there wasn't much open space inside them, so they didn't have to deal with the structural problems of modern skyscrapers.

To a certain extent the domes and spires of medieval churches and mosques did have to deal with those problems, but those too were primarily monumental and ceremonial, and while they contained lots of open space, they didn't contain much else. Some very clever engineering, and huge amounts of work (sometimes centuries of it for one edifice) went into finding ways to build those high roofs and keep them intact. But their builders did not take the next step of trying to convert that open space into places where large numbers of people could live and work.

The impetus to do that came in Chicago in the late nineteenth century.[9] A building boom following a disastrous fire in 1871 led to a great demand for property, especially for business in the downtown area near Lake Michigan and the Chicago River. Developers knew they could sell far more acreage than the land provided, if only they had it, and the only way to get it while also meeting the desire for "location, location, location" was to build high, stacking many acres of floor space above a single acre of ground.

But what were the limits? Up to then, a very tall office or apartment building meant four or five stories. Higher ones were impractical both because of the need to support extreme weights and because of limits on how much stair climbing people could or would do. As with the problem of airplane engines, the challenge of overcoming those limitations required the convergence of several lines of development, some obviously architectural, others less so. Some of the more important of those were new kinds of architectural design, metallurgy, electricity, and elevator design.

We might as well begin with elevators, since nobody would be interested in building very tall buildings if they couldn't get people, furnishings, and goods to and from the upper levels. Few people are willing to climb more than a few stories on a regular basis, and many couldn't if they wanted to. And it would be *very* hard to find people willing and able to carry a piano or a safe up, say, thirty stories.

That last consideration provided the impetus for creating the first elevators, and those go back at least as far as ancient Rome. At least as early as the first century BCE, Roman engineers used platforms on windlasses powered by water, animals, or slaves to lift freight. England in the early nineteenth century used similar devices with steam power; the lifting car itself was moved by ropes, cables, or hydraulics. The power needed was minimized by hanging the car on one end of a rope and a counterweight on the other, so that the motor (whatever it might be) had only to lift the load being carried and not the car itself.

Early elevators were strictly for freight. The ropes that carried the cars were too unreliable to trust with lives, so passenger elevators did not become a viable option until 1852. At that time Elisha Graves Otis, a builder and mechanic recently transferred to Yonkers, New York, to set up a new factory, invented the safety elevator—one that would not fall even if the supporting cable broke. The car slid on vertical guide rails, and in his new version, if the cable broke, clamps attached to the car automatically grabbed onto the rails, preventing it from moving.

Otis put his safety elevator to use in his current job, and the next year went into business for himself, building and selling them—but not nearly enough of them, until he applied a little showmanship. For a public demonstration in New York City in 1854, he rode one of his elevators to an impressive height and had the rope cut. Sales picked up somewhat after that—he installed the first safety elevator for passengers in a department store in New York City in 1856—but use of elevators took off dramatically when steam power gave way to electricity in the 1880s.

That change was fortuitously timed, because that was just when Chicago developers had their eyes on the sky, looking to build as high as they could. To do so, they needed not only elevators, but also new designs for the buildings themselves, and the key to those came from refinements in metallurgy. The earliest tall buildings were defined by load-bearing masonry walls, and the higher the building, the more weight the lower parts of the walls had to support. This meant they had to be extremely thick. Extremely thick walls meant less usable floor space, and required windows to be few, small, and not very effective at letting light and air in. But by then, it was possible to make large beams of cast iron or steel, extremely strong for their weight, and assemble those into a gridlike three-dimensional frame capable of withstanding the weight of the upper floors and their occupants, as well as the horizontal forces sometimes produced by winds or earthquakes.

Such frameworks could be built to unprecedented heights, and exterior walls, no longer having to support the weight of the building, could be quite light. In effect, metal-framed buildings just needed a thin "skin" (now commonly called a *curtain wall*) to protect the interior from weather and admit light. Eventually they would evolve to the type of construction often seen in recent decades, in which the exteriors of very large buildings consist almost entirely of windows. (See plate 5.)

There were still many engineering details to be worked out, and the higher the building, the greater the challenges. While steel frameworks can support a great deal of weight above them, ultimately the ground beneath the building has to support the whole structure. That's easiest in a place where bedrock lies near the surface, which may explain why skyscrapers first proliferated in New York rather than Chicago, where they were born. Many places don't have that advantage, and have to make do with relatively hard soil or extend solid supports to great depths to reach bedrock.

In the fifteen-story Reliance Building built in Chicago by William Hale in the early 1890s, for example, the vertical columns rested on firm clay, but still needed to distribute their weight over a wider area to prevent excessive sinking. This was achieved by adding a "spread footing," a vaguely pyramid-like structure of concrete, crisscrossed steel I-beams, and a cast-iron plate, at the base of each column. More sophisticated variations on this theme were developed to support even taller buildings. In some cases piles had to reach hundreds of feet down to touch or at least get close to bedrock.

Extremely tall buildings also need various forms of cross-bracing to stiffen their frames, not only because of their weight, but also because of the horizontal forces they must withstand due to wind (and, in some times and places, earthquakes). Even a very stiff building, hundreds of feet high, will sway noticeably, and even twist, in response to winds; and it must be able to do so without breaking. It's also desirable to keep the sway from getting too big in the first place, so some ingenious methods of doing this have been devised. In the Citigroup Center in New York City, for example, a "tuned mass damper" on the sixty-third floor uses a four-hundred-ton concrete slab mounted with springs and hydraulics as a huge shock absorber to keep the building as a whole from moving too much.

Sometimes the various requirements that a tall building must meet work at cross-purposes, and the builders must compromise. Building developers and owners would like to collect as much rent as possible, which means they'd like as much usable space as possible on every floor. Ideally, from their point of view, all exterior walls would go straight up, all the way to the top (as in the twin towers of the World Trade Center), but city governments have often insisted that upper stories be stepped back to let more light reach streets and sidewalks.

Adding more stories adds usable floor space, but elevators subtract from it—and the more floors you have, the more elevators you need to service them. Here, too, builders have used various design and traffic-control schemes to reach the best compromises, such as having different banks of elevators to serve different groups of floors, or having two cars "stacked" so that a single stop can let passengers on or off of two floors at once. But even with such ingenuity, the elevator problem lent itself to a continuing spiral. More floors required more elevators, more elevators required more floors to get as much usable space as the builder wanted, and so on.

Once those and other problems, such as ventilation, heating, and air-conditioning, were solved, buildings grew taller and taller. The motivations remained at least partly practical and economic, but there was also a significant and arguably less rational element of competition and symbolism. Builders vied to reach farther into the sky than anyone else, often adding slender masts or spires at the top to squeeze every possible inch of height out of the available budget (as in the Chrysler Building and Empire State Building in New York and the Petronas Towers in Kuala Lumpur).

Cities encouraged such competition because having the tallest building, or a lot of them, was seen as enhancing the prestige of the city itself. A dramatic recent example of symbolism is the new Freedom Tower being built on the former site of the World Trade Center—and being built even taller, as an overt gesture of defiance. Only time will tell whether that is a wise move, for the same convergence of characteristics that made the World Trade Center disaster possible will still be with us.

SMOOTH AND TURBULENT CONVERGENCES

In the world of literal streams, the convergence of small currents to form a big one can be either beneficial or destructive. Many tiny streams merge to form the Ohio River and others the Missouri, and those combine to form the Mississippi, one of the most important transportation routes in the history of our country. In the arid Southwest, after a summer thunderstorm, hundreds of tiny rivulets can arise suddenly and combine to form a raging torrent, a flash flood that can destroy tents and buildings, and drown campers and livestock.

Most of us would see aviation as a major benefit resulting from the convergence of research and development in such fields as aerodynamics, metallurgy, automotive transport, electricity, and navigation. We would also view the ability to construct very tall buildings as a major benefit resulting from the confluence of work in architectural design, metallurgy and other materials science, electricity, elevator technology, and plumbing and heating.

But as the World Trade Center so dramatically and painfully demonstrated, a confluence of two powerful historical currents can bring deadly destruction as well as great benefits. Modern airplanes are valuable because they can carry large numbers of people over large distances in short times, but the act of doing so means that they are inherently very massive objects moving at high speed and loaded with highly flammable fluids—which are exactly the qualities needed for an effective guided missile. We don't normally think of planes as missiles because normally they are guided carefully to gentle landings in specially prepared places.

But if that same mass and energy is instead guided to a hard impact with anything, the result will be massive destruction.

Similarly, skyscrapers make it possible to get a great many people, and a great deal of productivity, into a very small area; but the act of doing so also makes all those people and all that productivity vulnerable to very rapid destruction by a single bomb—or a hijacked airplane.

That combination of concentrated power and concentrated vulnerability is something any society that uses both airplanes and skyscrapers will have to live with and come to terms with. We may find ways to do so. Increased airline security can reduce the likelihood of any plane having its purpose so perverted (but that comes at a cost, only part of which is financial). New defenses can make it less likely that if a plane is hijacked anyway, it can reach its target; and engineers are working even now on improving ways to minimize destruction and loss of life if a building suffers a major hit. (Physicist and science fiction writer David Brin recently hosted a think tank for a History Channel special in which he and a group of engineers and firefighters brainstormed ways to rescue people trapped by fires in very tall buildings. Their suggestions included a robotic crawler that could climb the outside of a building, deliver firemen and their equipment, extract trapped occupants, and transport them back down to safety.)[10]

Both airplanes and tall buildings have given us such massive advantages that we will probably continue to use both as long as we can.[11] But doing so will require an unprecedented level of continuous vigilance, and new ways of thinking about how to make sure they can coexist peacefully.

CHAPTER 3
NEW ARTS AND SCIENCES

People commonly think of the arts and sciences as quite separate and distinct, perhaps even (as in C. P. Snow's *The Two Cultures*) as two poles of a gaping dichotomy. Yet, ironically, the implementation of any art depends on technology, and the convergence of old arts and new technologies can lead not only to new ways of doing art, but also to new *kinds* of art.

We just saw, for example, how advances in mechanics and metallurgy led architects quite literally to new heights. But the impact of new technologies on architecture was not merely one of size. In recent years it has become possible to build not only big, but also "strange," too, at least by the standards of earlier generations.

Consider, for example, the building in plate 6. When I first saw this structure, both inside and out, I marveled at the complexity of the sweeping curves all around me, and wondered how anyone could ever have drawn up plans for such a thing. A great deal of our architecture has always consisted of straight lines and right angles, occasionally varied by 30° or 45° angles, with an occasional circle or circular arc thrown in for artistic contrast—and, rarely, something still more complicated like a spiral staircase. Part of the reason simple geometries have been so dominant is that they're far easier to imagine, diagram, measure, and construct than more complicated ones.

Yet now we're seeing more and more buildings like the Experience Music Project/Science Fiction Museum and Hall of Fame in Seattle, other equally unusual buildings by Frank Gehry, or Rafael Viñoly's Kimmel Music Center in Philadelphia, a concert hall likely to remind the listener of being inside an enormous cello. How are they done? The answer was made clear on a television documentary[1] about Gehry, a Canadian-American architect, and you can easily find more about the approach to such marvels by searching Web sites for Gehry's name and the words *computer-aided design* or CAD.[2] CAD is a technique increasingly used in architecture and many other forms of design, enabling precise numerical computation of things too difficult and time-consuming to do without electronic assistance.

From early in his career, Frank Gehry's work often started with small-scale models built in his studio, of structures he would *like* to build. A tabletop mockup can be put together with relative ease and little expense, and has the virtue that it can be looked at from various angles to make sure it's what the architect wants. If he's dissatisfied, it's cheap and easy to change it, or even to scrap it and start over. But even if he creates a model that thoroughly satisfies him aesthetically, many practical questions remain: Can it be built at full size? Will it remain standing in the long haul, even through storms and earthquakes? Can it be heated and air-conditioned evenly and affordably?

Scaling things up is not as simple as it sounds. Shapes that are robust and stable in the workshop may collapse when the dimensions are increased by ten or a hundred times. Consider also that the real thing will be built out of different materials. Calculations of loads and stresses for complicated shapes are far more demanding than those for simple rectangular frameworks, but with fast computers with large memories and versatile displays, they become doable. Such computers can even be integrated into the very beginning of the design process, bypassing the need for physical models altogether. An architect can now imagine a building like none ever seen before and use the computer to see what it will look

Figure 3. Person-sized spider or spider-sized person? Either way, it won't work; you can't simply scale things up or down without changing other things about them. If you made a spider fifty times longer, wider, and taller, without changing its shape, its legs would have 2,500 times the cross-sectional area and therefore 2,500 times the strength—but its weight would be 125,000 times as great. It couldn't support its own weight. If you made a person fifty times smaller in height, again without changing shape, surface area would be so great compared to body mass that heat would radiate away too fast for warm-bloodedness to be possible. (Copyright © Wolf Read. Used with permission.)

like from any angle. If he's satisfied with that, he can feed in the dimensions and the properties of the materials he'd like to use for the real thing, and find out whether it's stable or needs modifications, such as making beams or walls thicker or using different materials. And when all that is done, the computer can generate plans that construction workers can use on the site to turn the concept into reality—with confidence that it will actually work.

Whether it *should* work, or should be attempted, is another question that will probably be hotly debated whenever something strikingly new is undertaken. Many visitors find the Science Fiction Museum building fascinating and even beautiful; others consider it a garish eyesore that clashes with its surroundings. In general, when people become capable of doing things that couldn't be done before, some will try testing their limits and do things that posterity will decide were mistakes, while others will be seen as major steps forward. You can think of it as a kind of evolution, with new conditions leading to mutations, and natural selection deciding which ones will survive and stand the test of time.

And such processes can be expected to occur wherever art and new technology converge.

SOUR NOTES OR SWEET NEW HARMONIES?

Music, for example has a long history of new technologies being invented or adapted by musicians, resisted by critics, and ultimately embraced or rejected by the listening public (or various segments of it, and sometimes more than once). When César Franck wrote his *Symphony in D Minor*, for example, one critic remarked scornfully that it obviously couldn't be a *real* symphony because it included a prominent solo for the English horn, a newfangled offshoot of the oboe family.

Adolphe Saxe's saxophone, a sort of hybrid made by putting a clarinet mouthpiece on a brass instrument with a conical bore, was received with contempt by most of the European classical music establishment, and still appears only rarely in symphonic music. But in twentieth-century America, its distinctive timbre quickly became, and long remained, a mainstay of the new art called *jazz*.

Brass instruments, like trumpets and horns, originally had no valves and could play only a few of the principal notes in a single key. If a composer wanted to use them in a piece that changed keys, the player had to change instruments. The introduction of valves around 1830 made that unnecessary: a valved instrument could play anything within its range, in any key. That was such a huge advantage that valved brasses now completely dominate the field. But I'd be willing to bet that they met resistance when they were introduced, too. Even today there are ensembles and conductors who specialize in "original instruments," giving performances played entirely on instruments as similar as possible to those used by their composers, hundreds of years ago. But most contemporary players prefer to take advantage of the best design features now available, suspecting that there's a reason why they replaced the older forms—and that even the composers would have preferred them, had they been available.

In our time technological advances, especially in electronics and data processing, have continued to influence music in a multitude of big ways. The development of commercial sound recording had several impacts, not all of which could have been easily foreseen. The initial and predictable reaction of professional musicians was alarm: fear that easy availability of recorded music would reduce public interest in live performance, and consequently their opportunities for employment. What really happened was in some ways surprising: while some of what the musicians feared did happen, the new phenomenon opened up new doors. Small, specialized recording companies could expose the public to music that had not been performed much in recent decades, like that of the Baroque period. Once exposed to such pieces via records, people became interested in hearing them live—so what seemed like a threat to live music actually expanded its scope.

And new kinds of instruments arose. Adding microphones to soft-spoken instruments like guitars let them be heard in noisy venues, but that was just the beginning. With the output of a guitar going through an amplifier on its way to an audience, it didn't take long to realize that electronics could do more to the sound than just make it bigger. Players began to experiment with things like "fuzz boxes" that distorted the sound in variable ways, some of which audiences liked. Also, with electronic amplification, stringed instruments no longer needed all the sound boost they got from the big sound box on an acoustic guitar. They needed little more than places to put the strings, the player's hands, and the microphone; everything else could be done electronically. In today's popular music, electric guitars of various shapes and sizes, with lots of options for electronically tinkering with the quality and quantity of sound, have become the backbone of most bands.

This is not to say that they have replaced the older acoustic instruments. Rather, they have expanded the range of choices available. Some kinds of performance, such as string quartets, concert bands, and symphony orchestras, still rely on the special qualities of acoustic strings, brasses, woodwinds, and percussion instruments that have evolved over the centuries. But all those instruments have one thing in common: their sound production depends on certain kinds of mechanical systems with a strong natural tendency to vibrate in very specific ways. Such a system might be, for example, a violin string, the air column in a flute or tuba, an oboe's reed, a trumpeter's lips, or a stretched drumhead. Those natural modes of vibration give each instrument its characteristic sound or "color," and make possible the rich palate of a symphony orchestra or a pipe organ.

But there are only so many kinds of mechanical vibrators that have been found to produce pleasing sounds. Can we imagine other kinds of sound that can't be produced by any such system? Yes, we can, and now we often hear them. A good loudspeaker does *not* have any special preference for particular frequencies or modes of vibration. It simply translates into sound any electrical signal that is fed to it, and it's possible to create electronically signals of any shape you might like. Fuzz boxes were an early example of this, intentionally reshaping the sound of a guitar string into new sounds that could not be produced by any mechanical oscillator. Synthesizers, first brought to popular attention by Robert Moog around 1964,[3] are a new kind of instrument that does away with the guitar string or its equivalent altogether. Instead they use pure electronics to create completely new kinds of electrical signals that are then converted to sounds never before heard.

Some synthesizers are essentially laboratory instruments, originally used by experimenters—some of them scientists or engineers, others composers—first just trying to see what effects they could produce. Later they used those effects to create new forms of avant-garde music. Others have incorporated those effects into more conventional but still novel musical instruments. Looking much like pianos and played in a similar way, those instruments, commonly just called *keyboards*, have become another of the most important components of popular bands in the late twentieth and early twenty-first centuries. They let players use familiar techniques learned on pianos or organs to produce a much wider range of sounds, even creating completely new tonal qualities.

Most of those instruments now use digital electronics, and from there it was a relatively small step to using computers, rather than human players, to generate the music itself. Composers and arrangers can use computers in at least three distinct ways:

1. To facilitate the writing and publication of music to be played by conventional instruments.
2. To facilitate writing music *and* performing it, with the computer itself (or at least devices attached to it) *replacing* conventional instruments and players.
3. To do the entire process, with the computer itself acting as both composer and all necessary performers.

The first of those is something that the famous composers of the past might have killed for, at least figuratively. Writing a symphony, for example, involves first writing a *score*, a manuscript showing all of the many instrumental parts being played at the same time (so that, for a large orchestra, a whole score page might contain only a single line of music, but occupying twenty or even thirty staves). Typically several of those instruments (but not always the same ones) might be playing the same thing, though some of them (for convoluted historical reasons) might be written in different keys. Often passages are repeated: sometimes a figure of just a few notes repeated over and over, with little or no change; sometimes a very long passage repeated over and over. In the days of Beethoven or Verdi, all this had to be done by hand—and if you've ever seen the penmanship of some of these people, you might well wonder how accurately copyists could ever have extracted individual parts from their manuscript scores. And if you've seen some of the copyists' penmanship, you might wonder how the players could read the handwritten parts at the first rehearsals and performances.

Present-day composers and performers have been freed from most of those burdens. Composers can buy off-the-shelf software packages with which they can write notes quickly, easily, accurately, and legibly with a couple of keystrokes and mouse clicks. If a sequence of notes, or an entire section of a score, is repeated, the composer can write it once, then copy and paste it as often as necessary. If a passage appears again in a different key, it can be pasted as it first appeared, then selected and transposed all at once by the computer. When it's time to copy out individual parts, the computer can do that, too—and both score and parts will be so neat that conductor and players will have no trouble reading them. Indeed, amateur composers and arrangers now routinely produce music visually indistinguishable from that professionally published.

The next step in computer-aided creation of music is to bypass human performers. Instead of using the computer just to prepare sheet music to be read by singers or instrumentalists, it can be used to control synthesizers, electronically directing them to convert digital descriptions into actual sounds played through speakers or headphones. Depending on the sophistication of the programmer, composer, and the equipment used, the sounds produced can be anything from simple little unaccompanied melodies up to convincing imitations of full-scale orchestras and choruses. Some programs, even ones preloaded on basic computers for home use, even allow people to compose without learning to read and write music, by playing their ideas on a "real" instrument, or one simulated on the computer screen, and letting the computer translate the sounds into musical notation.

It isn't easy to generate an artificial sound that can fool a listener into thinking it's a real cymbal or soprano, but with enough skill, patience, and processing power, it can be done. It's also possible, as I've already suggested, to create sound qualities not possible for any "natural" instruments, so that part of the work of a computer-using composer can be the creation of really new *kinds* of sound. Some start from scratch to do this; others use "sampling," taking snippets from recordings of past performances by others, or sounds found in the environment such as barking dogs or police sirens, and manipulating them into new forms.

But then, maybe no human composer needs to be involved at all, at least directly. At least some of the craft of composition can be analyzed as a series of particular kinds of choices, and a computer can be programmed to make such choices. This possibility disturbs some people, who like to think of some activities as exclusively the province of humans. When experimenters first began to talk about programming computers to write music, many said smugly, "Well, they'll never be able to write *real* music." Yet as early as 1997, composer-programmer David Cope wrote a program called EMI (Experiments in Musical Intelligence) that few could deny did just that. In a "musical blind taste test" reported by the *New York Times* in November of that year,[4] an audience at the University of Oregon listened to three unidentified compositions played by the same pianist: one by Johann Sebastian Bach, one by a professional musicologist imitating Bach, and the third by EMI. Asked to identify the one by Bach, the audience picked the one by the computer.

Does such a result mean there's less to music than we like to think—or that there's more to something else? If we grant that the music created by EMI is real music, even comparable to that of one of the most famous composers, who gets the credit for creating it? Should we recognize the computer as an intelligent entity deserving credit for creating art? Or does all the credit go to Dr. Cope, because he wrote the program, even though he played no direct role in deciding which notes would be played? (Think carefully before you answer; if you believe that, you may have to similarly say that human composers and artists deserve no credit for their works either, and must attribute it all to whatever Being or natural processes made them what they are.)

These are some of the more esoteric of the kinds of questions we will have to confront as computers do more and more of the labor of creation. Other such questions are more prosaic: How does copyright law apply to a composer "sampling" another's work for use in his own? When is it acceptable for a producer of musicals to substitute a synthesizer for a pit orchestra, thereby forcing dozens of musicians to seek other employment?

And so on.

PRETTY AS A . . .

Another of the most ancient and enduring arts practiced by humans is the making of pictures: two-dimensional patterns, often but not necessarily representing objects in the artist's environment, made by applying pigments to solid surfaces: drawing and/or painting. The motivations for doing this have varied widely—decoration, worship, recordkeeping, and personal tribute, to name just a few—none of which will concern us here. My subject is materials and techniques, the things that any artist, from a Paleolithic cave painter in France twenty thousand years ago to a contemporary Hollywood animator, must use to get an artistic vision into tangible form that a viewer can see and react to.

As those examples suggest, materials and techniques have varied a lot. The first known paintings were made by applying naturally occurring pigments such as plant extracts and wood ash to the stone walls of caves to represent such subjects as hunters and their quarry. In the ensuing thousands of years, the evolution of graphic technology consisted largely of such areas of applied chemistry as developing better pigments, liquid mediums to carry them, surfaces to apply them to, and tools (such as brushes) to do the actual transfer. "Better," in this context, means such things as bright, pleasing colors; ease of application, rapidity of drying, and lasting a long time without fading, changing color, or disintegrating.

The arts themselves experience convergences and interactions like those we've seen in connection with science and technology. One of the most important technological developments in picture making resulted from one of those, a convergence of painting and theater in the early nineteenth century. Louis Daguerre was a French artist and inventor whose specialty was painting scenery for plays, trying to make them not only entertaining but also as convincing as possible. He knew that one of the most realistic ways to get a picture onto a wall was the *camera obscura* (Latin for "dark chamber"), a technological curiosity that had long enjoyed some popularity in Europe.

The camera obscura is simply a dark, windowless room with a pinhole in the wall which acts as a simple lens to project inside a sharp image of the scene outside. You can still find them in a few places, such as Edinburgh, charging admission for people to come inside and look at the view outside in an unaccustomed way. Sometimes, for greater brightness, the pinhole was replaced by an actual lens made of glass, which could admit more light but still bring it to a sharp focus. Daguerre wondered whether there might be some way to make such a projected image permanent, and invented one utilizing the fact that silver salts darken when exposed to light. He coated copper plates with an emulsion of such salts and the light parts of an image projected on the plate darkened; washing away the unaffected parts left an image that came to be known as *daguerreotype.* The process was messy and cumbersome and the image not very good, by our standards, but the whole idea was unprecedented in 1829.

Other people, including Samuel F. B. Morse, made improvements in the photographic process, such as using glass instead of copper plates. By the 1840s photography had become an important tool for astronomical research. It had also captured popular interest for the novelty value of having a picture "painted" by sunlight. An added attraction was that it could record a scene very accurately, without the distortions that might be introduced (sometimes intentionally, sometimes not) by a human painter. But as long as it required messy chemistry, utilizing an emulsion that had to be mixed and smeared on the spot and used right away, it remained something that everyone could admire but only a few dedicated professionals could do.

That changed dramatically when George Eastman, another American inventor born poor and lacking formal education, became interested in photography and realized there could be a substantial market for portable equipment that virtually anyone could use. His first important innovation, in 1878, was learning to mix the light-sensitive emulsion with gelatin so the plate could be coated in advance, allowed to dry, and used later. In 1884 he developed a kind of "film," with the sensitive gel on paper instead of glass. That made possible the Kodak camera (the name, like many modern trade names, was chosen for its sound rather than any meaning), which became the basis of an immensely profitable business that has endured to the present, though with many substantial changes.

The first Kodak cameras, for example, weighed a couple of pounds. The whole camera had to be sent to Rochester to get the film developed. By 1889 the paper support for the emulsion had been replaced by celluloid film, and in 1924 that gave way to the less flammable cellulose acetate. By then the film could be removed from the camera for processing, instead of sending the whole thing in to be developed. In 1947 Edwin H. Land patented the Polaroid process, incorporating the processing chemicals into the film package so it could "develop itself" on the spot. That enjoyed some popularity, but the process was a bit cumbersome. Until recently the dominant medium for photography was film that had to be developed in a darkroom.

By using more complicated emulsions, with multiple layers containing different dyes, color was added to photography and eventually dominated the field, both in flat prints and in transparencies to be projected at high enlargements.

The biggest revolution since Eastman's Kodak was the one we are now in the midst of: the convergence of photography and digital processing. This has led to a new kind of photography that has already become more popular than the types using film. We'll take a closer look at that later, but long before that, photography experienced another massively important convergence.

LIGHTS, CAMERA, . . .

Though motion pictures represent a grand convergence of multiple arts and multiple technologies, photography was not one of the first of those. The first moving pictures were animations of sorts, nineteenth-century toys with names like Zoetrope, Choreutoscope, and Phenakistiscope. All of them depended on an effect called *persistence of vision*. If the eye is exposed very briefly to a scene, it retains the image for a slightly longer time. If it is shown successive pictures of a moving object taken at short intervals, on the order of one-thirtieth of a second, it sees them not as a series of distinct still pictures, but as a smooth progression; each one merges into the next and gives the same psychological impression as watching the actual motion.

Modern movies depend on the same effect, but whereas they (usually) show photographs of real scenes, objects, and people, the nineteenth-century persistence-of-vision toys used such devices as a rotating wheel containing drawings of successive stages of motion. A mechanism caused the viewer to see those scenes in rapid succession. In other words, they were an early form of what we now call *animation*.

Making movies of live people in physically real surroundings would require using photography to make the pictures, but, as we've already seen, early photographic methods were far too cumbersome for the job. Either shooting or showing dozens of glass plates per second just wasn't practical. The biggest jump in solving that problem was the replacement of glass plates with Eastman's celluloid film, which could be wound on spools and driven through a viewer or projector. In this country, Thomas Edison usually gets the credit for that, even though most of the actual work on the kinetograph (camera) and Kinetoscope (viewer) he "invented" in the late 1880s was done by an employee of his named William Kennedy Laurie Dickson. Also, Edison missed the importance of projecting the pictures for a large audience. He saw them primarily as an auxiliary accompaniment to his sound recordings, and his viewers were "peep-show" devices for a single user. The pioneering work on projection was done by others in Germany, England, and France.

If you consider Edison's view of moving pictures as an adjunct to sound, it's ironic that the first movies to enjoy popularity were silent. More precisely, they didn't include dialogue, though they were usually accompanied in theaters by live musicians—sometimes a piano, sometimes a small ensemble—who played music to suit the action. Two major reasons for this were the difficulty of synchronizing recorded sound with the pictures, and the challenge of making the sound loud enough to be heard by everyone in a theater.

The solutions came from electronics. Lee De Forest's triode amplifier, mentioned earlier, made it possible to convert sound to a big enough electrical signal to drive a loudspeaker. Such a signal could also be displayed visually on an oscilloscope, an instrument invented in 1897 in which an electron beam "jiggled" by sound waves traced a graph on a fluorescent screen. That led to a technique for recording sound as a photographic trace along the edge of the same film containing the pictures, which put a decisive end to the synchronization problem.

And so movies became talkies. *The Jazz Singer* (1927), starring Al Jolson, was the first commercial success with this new fusion of techniques, and led to a vast number of others. Silent movies became a thing of the past (though not without some lamenting the loss of their unique ability to transcend language barriers). Audiences came to take for granted that movies would integrate visual storytelling with music and dialogue. The field grew, flourished, and proliferated into a vast array of types and subtypes. Technology and art continued to interact, spawning new techniques such as fades, dissolves, zooms and pans, and various kinds of trick photography. Color became the rule rather than the exception; in *The Wizard of Oz* (1939), Dorothy's transport from prosaic Kansas to magical Oz was underlined by the sudden blossoming from black-and-white to color. Still later, there would be new techniques, and new artistic uses for them, such as 3-D, IMAX, and IMAX 3-D. Movie actors, perhaps even more than stage actors before them, became huge icons of popular culture.

But something else happened concurrently. Remember that the earliest moving pictures were not photographed, but drawn. Even as movies developed more and more ability to show realistically the actions of live people and animals, some of them exploited instead the ability of animation to show events that could never happen in reality. Coyotes in pursuit of roadrunners could run straight off cliffs, stop instantly in midair with legs churning until they realized where they were, and then plummet straight down. In the early days, animation was most often used in just such short cartoons, shown before the main feature in a theater. There were occasional feature-length animations, such as Walt Disney's *Fantasia* (1940). But while some of those were extremely clever and imaginative in concept and execution, it was always obvious that they were animations. Even at their best, the pictures had a simplified, stylized quality that no one would ever mistake for reality.

Recently this has changed dramatically. Here again the convergence of computers with older technologies has led to a qualitative leap in what can be done, and therefore what is done. Traditional animation is a laborious, time-consuming process, requiring each frame—and remember, that means twenty or thirty of them for every *second* of screen time—to be created individually. Usually this animation was done by drawing, painting, or making three-dimensional models whose parts could be repositioned, a little at a time, for successive shots. Modern computers, with high processing speeds and the ability to store and manipulate huge quantities of data, have automated a great deal of this detail work.

For example, suppose the first frame of a sequence is drawn in detail. If the computer is programmed with information about how a figure in the scene will move, it can calculate, quickly and without direct human oversight, how each successive frame will differ from the one before it, and "draw" all of them. The drawing process itself can be very detailed, including subtle color shadings and lighting effects that no human animator would have time to do by hand.[5] The computer can even automatically determine the brightness of each part of the picture, and where and how deep the shadows should be, if given information about the size and shapes of the objects in the scene and the location, brightness, and color of the light source—even if the light source is moving or changing with time.

The most obvious consequence of computer animation is a powerful blurring of the line between animation and live action in what the moviegoer sees. Feature-length animated movies are now being produced (such as *Happy Feet*, the 2006 fantasy about penguins in Antarctica) in which many scenes could easily be passed off as high-quality photographs. But this narrowing of the gap isn't limited to animated movies. The techniques of highly realistic animation are being incorporated into movies that are predominantly live action. In the *Lord of the Rings* movies, for example, the story is basically one about, and performed by, real actors. But some of the more fantastic characters, such as Gollum, and even humans not seen too closely (sometimes thousands of them in battle scenes), are entirely computer-generated. Yet they are so minutely detailed that the entire scene looks photographed, and it takes close scrutiny by a skilled eye to tell which parts are real and which are not.

For many years, special effects in science fiction and fantasy movies were crude and easy to recognize. For many moviegoers, really good special effects—those giving a convincing illusion of reality, without the "strings" showing—were so rare that they were sufficient reason to see and enjoy a film. Too many movies relied heavily on that fact, drawing viewers in with spectacular effects that made the audience overlook weak stories and wooden acting. Those days may well be ending. So many studios now routinely produce effects that would have been impossible a few decades ago that audiences are beginning to take those effects for granted. They are even used in commercials, and fantasy movies may finally have to rely for their competitive edge on strong stories and performances.

How far can the trend go? The hardest thing to simulate convincingly by computer animation is the human face and body. Crowd scenes are not too formidable, because nobody in them is seen in intimate detail. Close-ups are far more challenging. Since much of human life is devoted to interaction with other humans, our nervous systems are hard-wired to be extremely sensitive to the details of how people look and behave. Fooling people with simulations of those details is far from easy, yet even there, the principles are becoming ever better understood and the efforts growing more successful. Science fiction writers have imagined futures in which human actors and actresses have been largely or completely replaced by computer simulations (see, for example, "The Ghost in the Machine," by Grey Rollins).[6] Such a future may or may not be good for the public (probably, like most things, it would be a mixed blessing), but the possibility is far less remote than actors might like to believe.

ENTER THE AUDIENCE, STAGE LEFT

There is one more cluster of major changes resulting from the convergence of technology and art that is already very much with us: the integration of the reader or viewer into the art as an active participant. Earlier I mentioned hypertext as a new way of organizing information, allowing such features as the ability to jump instantaneously from one part of the text to another, to link publications directly to critical commentary on them, and so forth. Such a system has obvious advantages as a reference tool, and there are now several online encyclopedias, including Wikipedia,[7] which is continually modified by its users. The internet itself, in its entirety, could well be viewed as the world's largest (and worst-edited) encyclopedia.

There have also been experiments with telling stories in hypertext form. Marc Stiegler's *David's Sling*,[8] for example, was first told in 1988 as a conventional (at least in form) novel: a narration of events chosen and arranged by the author as a book to be read straight through from beginning to end. But at the same time, Stiegler (himself an information technologist and hypermedia pioneer) published a hypertext version of the story allowing the reader such options as reading the story from different points of view and jumping from the text into discussions about the underlying ideas. This disk version also included reference materials on the characters, settings, and technology involved in the story—illustrated with pictures and animation.

That's why I referred to Stiegler as a pioneer in hypermedia, not just hypertext. Originally hypertext was mainly a new way to organize words; but as the ability to handle other kinds of information digitally grew, the boundaries between words, pictures, and music dissolved. In just a few years, most of us have come to take hypermedia for granted as a pervasive part of our everyday environment. Children are growing up as much at home in cyberspace as in the physical world, jumping freely from site to site via links activated by clicking a mouse. They take it for granted that each new jump will take them not only to new information, but also to new sources of entertainment laced with lively graphics, animation, and sound effects.

In a similar vein, interactive computer games are the norm for young people today. We sometimes hear that such games are ruining the younger generation, but it is not at all clear that this is the case. What is clear is that in an evolutionary sense, they carry all the earlier convergences to a new level, not only merging the visual and other arts into something quite new, but also allowing the player to become an active participant in an elaborate story scenario. Some of their antecedents were interactive novels that offered options like those in *David's Sling*, mentioned above, plus others that gave readers the opportunity to choose one of several possible endings. The most elaborate games, however, go further: the writer/programmer creates a complex, detailed scenario in which stories can take place—and then makes the user an active participant in the story, *becoming* a character and making major decisions about how the story will actually develop.

This is unprecedented, but is it in any real sense "the culmination" of those convergences? Probably not. People have long tended to assume that they are the ultimate product of creation (or evolution), and their current world the climax of history. So far, at least, they have always been wrong: the future has always held plenty of surprises. I see no reason to suppose that our future will be any less so.

As a sobering reminder of just how true that is likely to be, let me conclude this chapter by quoting John W. Campbell's introduction to Albert W. Kuhfeld's fact article on "Spacewar" in the July 1971 *Analog*.[9] Spacewar was perhaps the first interactive video game, in which players controlling blips on a computer screen, representing spaceships in a solar system with simulated gravity, tried to blow one another up. It was originally developed by students at MIT; I had a chance to try it out while visiting one of my graduate school advisers after he relocated to MIT. As the first thing of its kind, it was really something; it made jaws drop, and hooked people into playing it and modifying it for hours on end. By the standards of what we're now used to, it was, purely and simply, quite primitive.

In 1971 Campbell wrote of it, "It's a great game, involving genuine skill . . . but I'm afraid it will never be widely popular. The playing 'board' costs about a quarter of a megabuck!"

Less than a decade later, my nine-year-old nephew received a far more sophisticated descendant of it as one of several Christmas presents, to be played on any handy television set. And look where we are now: far more sophisticated descendants of *that* now *are* wildly popular.

And the convergences continue to come, faster than ever.

CHAPTER 4
LOOKING INSIDE
New Technologies and Medicine

Earlier I promised to take a closer look at how a CT scan works, and why it took a convergence of technologies to make it possible. Before we can appreciate the technologies to come, let's look at a much simpler example of something that illustrates the same basic principle as a CT scan.

You may remember that its basic idea is to get a three-dimensional picture of the complete interior of some body part such as the head or chest. An ordinary x-ray doesn't do that; it gives an immediately recognizable picture, but it's a simple silhouette that just shows how much radiation made it through the body along the line from the x-ray source to each point on the film. It tells nothing about how the absorption varied along that path. Was it absorbed at a constant rate all the way through; or did it pass easily through some parts and get absorbed mostly by one compact, densely packed mass like a bone or tumor?

A CT scan doesn't give a picture directly, but instead collects numbers with which a computer can calculate and "draw" a picture showing how the density of matter varies throughout a "slice" of the body or head. If you collect enough such slices, together they add up to a three-dimensional picture of the entire head or body.

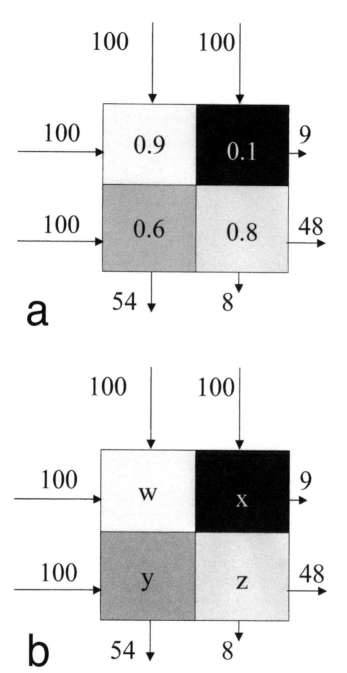

Figure 4 (a&b). A simple made-up example to illustrate the principle of a CT scan. In each case, we're looking end-on at four parallel bars of different partially transparent materials. In (a), we know how much light is transmitted by each material, so we can easily calculate how much light passes through each pair of bars. In (b), we know only that there are four bars, and how much light passes through the combination. From that we must figure out the four unknown transparencies. (Diagram by the author.)

The secret to constructing the picture of each cross-sectional slice is to pass x-rays through the patient in many different directions, measure what comes out the other side, and figure out what distribution of mass would produce those results. Let's see what's involved in doing that.

Figure 4a shows a cross-section of a bundle of four bars laid side by side, each made of a different partially transparent material—a very simplified stand-in for a human patient. Suppose we shine a light straight at it, first from the left and then from the top. (Visible light is the same kind of stuff as x-rays, but made of longer waves—and easier for most of us to visualize.) The intensity, or amount of light hitting the surface, is adjusted to be one hundred arbitrary units. Each bar is shaded to suggest how transparent or opaque it is; the number at its center is the fraction of the light hitting one face that makes it through to the opposite face. If you know those fractions, it's easy to figure out how much light goes through each part of the entire bundle. For light passing straight through the top pair of bars, for example, 90 percent makes it through the left-hand bar, but only 10 percent of that makes it through the right-hand bar—so only nine units of light come out the other side of the bundle. Similarly, for the lower pair of bars, the amount coming out is $100 \times 0.6 \times 0.8 = 48$, and for the left and right vertical pairs the amounts getting through are fifty-four and eight.

In medicine the problem is turned around. In figure 4b, we have the same situation but we don't start off knowing how much light is transmitted by each bar. Instead, we're trying to determine the transmitted fractions, which we identify as the unknown quantities w, x, y, and z, by shining a one-hundred-unit light on the left and top faces and measuring what comes out in the places indicated. If those values are the ones we just calculated, then our problem is to find the values of w, x, y, and z that will make all these statements true at the same time: $100wx = 9$, $100yz = 48$, $100wy = 54$, and $100xz = 8$. In other words, we have four equations to solve simultaneously for four unknowns, and they're all intertwined, with two of the unknowns in each of the four equations.

This is a bit scary. In my first example, we just had four equations, each with only one unknown, and figuring out the unknown quantity required only simple arithmetic. Even two equations with two unknowns are usually pretty simple. For example, if we want to find x and y such that $y = x + 3$ and $2x + y = 30$, we can just replace y in the second equation by $x + 3$, turning that equation into $2x + x + 3 = 30$. Therefore $3x = 27$, $x = 9$, and $y = x + 3$, or 12.

If you don't use math much, you may be thinking this is about as far as you care to go with this sort of thing, and indeed the difficulty of solving multiple simultaneous equations with multiple unknowns goes up rapidly as the numbers of equations and unknowns go up. Even for four, as in my contrived example, it's hard enough that I'm not going to give the details here. Mathematicians have ways of doing it, even for much larger numbers, but the amount of number crunching required becomes horrendous; it would take so long it would be impractical for anybody to do it by hand.

That's why *computer* is the first part of Computerized Tomography. The clinical problem for a patient undergoing a CT scan is similar in prin-

Figure 5. The bare essentials of a typical setup for a real CT scan. (Diagram by the author.)

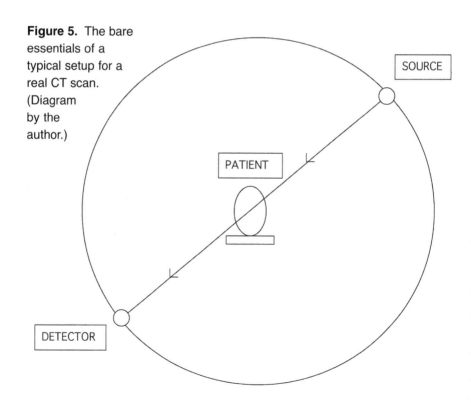

ciple to my figure 4b, except that a patient is far more complicated than a set of four uniform parallel bars. Figure 5 shows a typical CT scan setup, with the patient lying on a table.

The ellipse in the middle represents her head, viewed from the top. An x-ray source is mounted so it can be positioned anywhere around a circle surrounding the patient, and an x-ray detector is moved so it is always at the opposite end of a line through the patient. In practice, the source and detector are moved simultaneously, and fast, so that a complete scan (collection of readings) is taken in just a few seconds. The computer then analyzes the results by dividing the patient's cross-section into a large number of small regions (like the four bars in my opening examples, except that there are a lot more of them) and figuring out what the transmission by each little region must be to produce the observed results. If each region in a map of the cross-section is shaded to show its calculated transmission, the result is a picture like figure 6. This is a view that could not be obtained by any single x-ray photograph, and allows

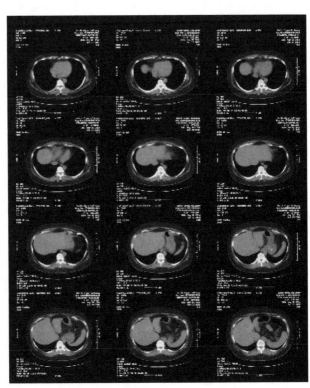

Figure 6.
A typical CT scan of a healthy human torso. Each picture in the set shows a "slice," so that all of them taken together give the doctor a three-dimensional picture of the internal organs. (Reproduced by permission of Henry G. Stratmann, MD, and Maryellen Stratmann, MD.)

precise pinpointing of any feature—such as a tumor—in the patient's body.

And that is valuable because it tells a surgeon exactly where something needs to be fixed—even if the lesion, when exposed, is not visually recognizable as such.

VARIATIONS ON THE THEME: OTHER KINDS OF TOMOGRAPHY

The same principle of using computer analysis to see things inside the body that would not be apparent in any single view has also been applied to other types of medical imaging.[1] Why do we need more than one kind? Because each has particular strengths—things that it does well that the others can't do as well (or at all).

X-rays for example, first found medical favor because they passed easily through soft tissues but not through dense materials such as bones and teeth. This made them excellent for seeing things like fractures, bone deformities, and dental cavities, and to a lesser extent softer parts that absorbed the rays better than surrounding parts. But the contrast associated with such slight variations was low, and opaque parts like bones obstructed the view of softer parts in line with them. So what could you do if you needed a good, sharp, high-contrast view of details within soft tissues like muscles or lungs?

A useful technique for those is *magnetic resonance imaging*. The full name of the effect that this technique depends on is *nuclear magnetic resonance* (NMR), but it has nothing to do with radioactivity. (I suspect the *nuclear* has been dropped in common usage to avoid the negative reaction many people automatically have to the word.) Bulk matter is made of atoms, each of which consists of a tiny nucleus, with a positive electrical charge, surrounded by a cloud of negative electrons. The nucleus is much more massive than all the electrons put together—in other words, it contains almost the entire mass of the atom—and among other things it acts like a magnet. That means that if you put a collection of matter, be it a rock or a patient, in a strong magnetic field, all those little magnets will try to line up with the field lines.

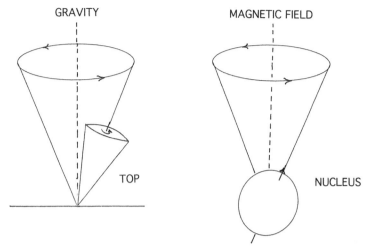

Figure 7. Precession in gravitational and magnetic fields. A spinning top or gyroscope with its lower tip fixed, as at left, will slowly move so that its spin axis traces out a cone around a vertical line. An atomic nucleus acts like a tiny magnet along its axis, and that axis undergoes the same kind of motion around a line of the magnetic field, as at right. (Diagram by the author.)

For esoteric reasons that deal with quantum mechanics, the nuclear magnets can't simply line up exactly with the magnetic field and stand there at attention. Instead, they're tilted at a precisely determined angle and precess around it, just like a spinning gyroscope with one end on a tabletop or a pedestal (see figure 7). In the case of a nucleus in a magnetic field, the frequency (rate) of this motion (sometimes imprecisely described as a "wobble") is determined by properties of the nucleus (mass—and magnetic moment, which means, roughly, a measure of how strong a magnet it is) and the strength of the applied magnetic field.

Radio waves, like light or x-rays, consist of periodically varying electric and magnetic fields. If a quick burst of radio waves, varying at the same frequency as the precession rate, is beamed at a collection of magnetically aligned nuclei, it can knock them out of alignment, much as a loose windowpane might vibrate if you make a loud sound of a certain pitch near it (hence the word *resonance*). When the radio pulse ends, the atoms again line up to precess around field lines at the characteristic frequency, emitting in the process their own burst of radio waves at that frequency.

In medical magnetic resonance imaging, or MRI, the patient is surrounded by powerful electromagnets (themselves the product of convergence of several technologies including solid-state physics and the production of very low temperatures). By activating different magnets at different times, a computer-assisted technician can examine the magnetic resonance effect in different parts of the body, subjecting each one to a known magnetic field and bursts of radio waves and measuring the answering waves. The computer can compile the results, much as in CT scanning, to yield an image of how the resonance response varies from place to place within a slice of the patient's body.

The special value of this for medicine depends on the choice of the nuclei used for the resonance. The usual choice is hydrogen, which is very abundant in the body and shows a strong nuclear magnetic resonance. Overall, water makes up about 70 percent of the body, but just how abundant it is varies from one kind of tissue to another. The resonance effect will be strongest in parts of the body that contain the highest percentage of water (and therefore hydrogen). So measuring how the resonance strength varies from region to region allows the construction of high-contrast images of soft tissue (which is hard to do with x-rays because they mostly pass right through it). Best of all, teeth and bones—the very parts that are nearly opaque to x-rays—contain very little water and so show up faintly, if at all, in MRI. Thus MRI can be used even for tough jobs like examining the spinal cord, which consists of soft tissue encased in bone, and is therefore effectively hidden from x-rays.

The upshot is that MRI makes an excellent complement to CT: one is excellent for detailed examination of soft tissues, the other for hard.

Several other kinds of hybrid imaging technologies have helped revolutionize medicine in recent years. I'll mention just one more: the PET scan, or *positron emission tomography*. This one does involve radioactivity, and of a particularly dramatic sort (yet is quite safe when used correctly). The PET scan has the special virtue of giving not only a detailed three-dimensional picture of the patient's interior, but also one with a time element: it shows not only what's located at a particular point in the body, but also what's happening there at a particular time.[2] The key to it is a radioactive tracer: a radioactive version of an element that can be attached to something normally carried through the body by the circulatory system. Seeing where the radiation is tells the doctor something about active processes in particular parts of the body, such as circulation, metabolism, thinking, and feeling.

As I mentioned earlier, every atom consists of a positively charged nucleus surrounded by negative electrons. A particular element, such as oxygen or carbon, is characterized by a particular number of positive protons in the nucleus and an equal number of negative electrons, which determines how it behaves chemically. In addition to protons, the nucleus contains particles called neutrons, very similar to protons except that they have no electrical charge. Atoms of any element can occur in several slightly different forms called *isotopes*, differing only in the number of neutrons in the nucleus.

Since chemical properties are determined only by the number of electrons, different isotopes of, say, oxygen or carbon are interchangeable in chemical processes such as metabolism. But they differ in other ways that make them useful. Usually an element consists almost entirely of one isotope because that one is stable, while other isotopes are radioactive. This means changes occur spontaneously that cause the nucleus to change into a different kind of nucleus. For example, almost all nitrogen, which plays an important role in biochemistry, is nitrogen-14, a stable isotope with seven protons and seven neutrons. Nitrogen-13, with only six neutrons, does not occur naturally because it spontaneously (and quickly) decays into the stable but rare carbon-13. This happens when one of its protons changes abruptly to a neutron by spitting out a *positron*, a particle just like an electron except that it has a positive charge.

And when a positron encounters an ordinary negative electron, they destroy each other instantly, converting all their mass into energy that's carried off by two gamma rays (yet another form of electromagnetic radiation) in opposite directions. Since the two rays are produced simultaneously at the same place, and travel in opposite directions, you can tell where they were emitted by setting up a bunch of gamma ray detectors and electronic circuitry to record which pairs of detectors were triggered simultaneously. Each annihilation event occurs somewhere near the middle of the line between the two detectors it triggers, so many such lines pass through regions where many positrons are being produced— which are where the body is concentrating the tracer isotope.

Nuclear physicists have developed ways to produce short-lived radioisotopes such as nitrogen-13 in machines called *accelerators*, because they accelerate elementary particles to high speeds and let them slam into target atoms, thus changing them to new isotopes. Half of any nitrogen-13 nuclei that you have will decay in about ten minutes. But if you act quickly you can make nitrogen-13 atoms, attach them to a suitable biologically active chemical, inject them into a patient, and use a bank of gamma ray detectors as described above to keep a record of where positron-electron annihilation is taking place. That information lets the computer construct an image showing where the radioactively tagged chemicals are going and how their distribution changes with time. That gives doctors an unprecedented ability to determine such things as which parts of the heart are receiving an adequate supply of blood and which are not, or which parts of the brain are most actively involved in solving a calculus problem or writing a sonnet.

WHAT DOES IT MEAN (AND WHERE HAVE I SEEN THAT BEFORE?)

You may well deduce from my descriptions of CT, MRI, and PET that a goodly amount of skill and experience is necessary to know what the images actually mean. The human body and mind are extremely complicated physical and chemical systems, and a scan using one of those methods may give a pattern not quite unlike anything the doctor has ever seen before. It will be only one of several kinds of information he or she needs in order to make a diagnosis and decide on a treatment. One thing that can help to make the best use of that information is the ability to compare it to as many as possible previous patterns, whether seen by the same doctor or others, to determine which ones looked most like this pattern and what did or didn't work in those cases.

That's yet another way that computers can help doctors: not just by creating new kinds of diagnostic images, but comparing them to large databases of similar images collected by many doctors in the past. That kind of help isn't limited to sophisticated ultramodern imaging techniques. Sometimes it can be a powerful tool even in kinds of diagnosis that have traditionally depended on the doctor's personal skills of observation and judgment. Some of those don't even involve "looking inside" at all.

Take skin cancer, for example: right out in plain view, and typically diagnosed by visual inspection, followed by a biopsy if the doctor considers it necessary. There are three main types of skin cancer. Two of them, basal cell carcinoma and squamous cell carcinoma, are relatively nonthreatening. They don't usually metastasize, or spread to remote parts of the body, and can usually be completely cured if they're caught at a reasonably early stage.

The third type, malignant melanoma, is another matter entirely. It grows fast, metastasizes easily, and until pretty recently killed most people who got it. In recent years the cure rate has been improved, but the cure depends on cutting out the entire cancer and some of the surrounding tissue *before* it metastasizes. The difficulty is diagnosis: melanomas vary greatly in appearance and can look very similar to some benign growths. I once asked a dermatologist if there was a way I could tell which blemishes were harmless and which needed his attention and possibly a biopsy, and he said, "Sure, do a six-year residency in dermatology."

Even doctors who have done lengthy residencies sometimes get it wrong, because nobody has enough experience to have seen all possible variations of melanomas and the harmless growths that can be confused with them. But what if every doctor could draw on the experience of all doctors, comparing what he or she sees with what everybody has seen? Early in 2007 one company announced a way to do something very much like that.[3]

MelaFind is a handheld scanning device that looks a little like a hair dryer and incorporates a camera, an electronic database, and computer software for quickly comparing what the camera sees to thousands of images in the database showing different kinds of melanomas and lookalikes. Preliminary tests show that the instrument catches many more melanomas, and recommends fewer unnecessary biopsies, than even an excellent dermatologist relying only on his or her eyes. It, or something like it, seems likely to become a standard tool not only for melanoma recognition but also for many other kinds of diagnosis that depend on recognizing pathological patterns.

LOOK, BUT ~~DON'T~~ LIGHTLY TOUCH

One more category of devices for looking inside the body depends not on esoteric analysis of data derived from processes like NMR or positron emission and annihilation, but simply on getting ordinary vision into places it normally couldn't reach. Those devices, too, depend on converging technologies, and in some cases, they also open up the option of *doing* things inside the body less invasively than before.

Some of the most familiar and important of those devices are those used for *endoscopy*, a term that simply means "looking inside." The idea isn't entirely new; doctors have used some forms of it for quite a while to look at organs accessible through natural openings (such as the esophagus, stomach, or bladder) or through an incision made for the purpose (such as the lungs or liver). Relatively few body parts lend themselves to this kind of examination, and the process has been at best inconvenient and uncomfortable for the patient. The lighted rod or tube used was rigid or semirigid, and getting it in and out typically required anesthesia and/or hospitalization. The general principle, however, can now be applied to more organs, including some deeper inside the body and/or otherwise hard to reach, and with less trauma for the patient, thanks to a relatively new technology called *fiber optics.*

The key concept here is *total internal reflection.* Normally, if light traveling through a dense transparent medium like glass or water comes to an interface with a less dense medium like air, part of the light is refracted (transmitted but bent), and part of it is reflected back inside the denser medium. How much is reflected and how much is refracted depends on the angle at which the light hits the interface. If it hits straight on, it all goes straight through; if it skims along the surface, none goes through. Anytime the light hits the interface at a sufficiently glancing angle, none of it goes beyond; it all stays inside the denser medium.

This total internal reflection makes possible the new and very useful category of fiber optics. If you shine light into the end of a thin glass fiber, it will travel all the way through to the other end, bouncing off the outer surface whenever it touches it, with no loss of brightness. This happens even if the fiber is quite long and curved or twisted. This effect can be useful if, for example, you need to get light into a hard-to-reach part of a machine you're inspecting. If you just used a flashlight, some parts would be shaded by others. But if you mount the flashlight on one end of a fiber optic bundle, and twist the bundle into a shape that you can snake around the obstructions, you're in business.

If you bundle a large number of fibers carefully, keeping them parallel so that they're in the same relative positions at both ends, you can do something even better: send an image through a fiber optic "light pipe" so that an observer at one end of the cable sees what the other end sees, perhaps in a place that the observer can't see directly.

Medical endoscopy typically uses both those capabilities at once. If two light pipes are threaded into a body cavity side by side, one of them can act as an illuminator while the other lets a doctor look at what's being illuminated—say, a possibly precancerous intestinal polyp. Before fiber optics, this could be done only near the lower end of the intestine. But a fiber optic endoscope can go as far, and around as many twists and turns, as necessary. Thanks to still other converging technologies, those of miniaturization and remote-controlled manipulators, it's now routine for the examining cable to carry along a tool for snipping out suspicious-looking growths for microscopic examination, without requiring a separate surgical procedure after the endoscopy.

On the other hand, sometimes a separate surgery will be necessary. One drawback of a routine colonoscopy done as I've described is that, in general, the doctor depends on recognizing potential problems in real time during the brief period when the equipment is in place. As I mentioned earlier in connection with melanoma, it's possible for even a skilled practitioner to overlook something, or not recognize its significance right away. A doctor using an endoscope may recognize something as *possibly* cancerous, but be unsure whether it is or not until it's been removed and sent to a lab. If it turns out to be benign, no further action may be needed; but if it's malignant, major surgery may be needed right away. Nobody wants that unless it's truly necessary. So electronic means have recently been developed for storing endoscopic images to allow a more unhurried visual examination than a real-time procedure allows.

It's also possible to do the whole thing electronically. Instead of inserting a cable containing fiber optics for lighting and viewing, one could thread in a cable containing a light source and a small camera for electronically transmitting a television picture of an internal organ. Given current levels of miniaturization, this is possible; but for many purposes fiber optics have the virtue of simplicity: fewer interacting parts, so less can go wrong.

There is, however, one particularly intriguing version of the remote-camera method that has recently shown considerable promise. Even though fiber optics makes endoscopy more flexible (no pun intended) and less intrusive, nobody considers it fun. Recently researchers have developed a tiny television camera that the patient can actually swallow, like a pill, and wait while it sends out pictures of the digestive tract as it passes through.[4] This is surely an excellent example of a welcome technique that medical practitioners would not have developed without the help of other fields.

One more bright note for patients: the same trick I mentioned in connection with colonoscopy, sending in tiny, remotely controlled surgical tools along with the examining cable, has opened up (again, no pun intended) a whole new field called laparoscopic surgery. Procedures that just a few years ago would have entailed major incisions and recovery times of weeks or months can now be done with minimally invasive methods in which remotely controlled tools and optical devices are inserted through quite small incisions, and the patient recovers in days instead of weeks.

And who could complain about that?

CHAPTER 5
COMPUTERS AND GENES

As we have seen, our contemporary civilization has been shaped in very large measure by the emergence of systems for handling large quantities of information. This influence promises to be even greater in the future. However, not just our local culture, but the entirety of life on Earth, has been shaped by an even larger and far more ancient such system. Only in the last few decades have we begun to understand clearly how it works; and now that we do, that understanding promises to have an enormous impact on how we live in the future.

Our recently won understanding depends, like so many other things in today's world, on a convergence of several kinds of science and technology, and its progress was not smooth or uniform. It began in 1857, with the Augustinian monk Gregor Johann Mendel, who grew peas in a monastery garden in what is now Brno, in the Czech Republic. Earlier he had studied and taught math and science, though not at the schools he would have preferred. And still earlier, during a poor childhood, he had tended fruit trees. The monastery garden gave him a chance to combine his mathematical and botanical interests in research that he conducted, likely with little idea of the great importance it would eventually have.

Before Mendel's time people had some awareness of inherited characteristics, but no specific understanding of how the process worked. Darwin had his theory of how natural selection caused favorable variations (mutations) to proliferate and detrimental ones to die out. But he was vague on how variations arose in the first place, and troubled by the notion that new and old characteristics might blend to some intermediate state before natural selection had a chance to work. Mendel's great contribution was the demonstration that traits did *not* blend, but remained distinct and were inherited according to rather simple empirical rules.

Mendel's pea plants, for example, could be classified into two groups: "tall" and "dwarf." Whether plants were self-pollinated or cross-pollinated, the offspring still fell into the same easily recognizable categories. But the relative numbers of offspring in each category depended on their parentage. Dwarf plants always produced dwarf plants. About a third of his tall plants produced only tall offspring—but the rest produced sometimes tall and sometimes dwarf, with a ratio of about three to one.

All this could be explained, or at least systematically described (which is not quite the same thing), by assuming that a plant's tall or dwarf nature was determined by something called *genes* carried in seeds and pollen. Each parent contributed one gene, which could be either tall or dwarf, and how the offspring developed depended on what kind of gene pair it got— and the fact that the tall gene was "dominant" and the dwarf gene recessive. *Dominant* meant that if even one gene of the pair was tall, the plant would be tall, regardless of what the other gene was. *Recessive* meant that the dwarf trait would assert itself only if both genes of the pair were dwarf. Dwarf plants always produced dwarf offspring because, to be dwarfed, they had and could pass on only dwarf genes. Tall plants with both tall genes would produce only tall offspring, because they had only tall genes to pass on. But tall plants with one tall and one dwarf gene could pass on either kind, and their offspring could be either tall or dwarf, depending on which gene they acquired from the other parent. Furthermore, genes came in "packages" called *chromosomes*, so that all the genes on one chromosome were passed on together and inseparably.

This is essentially how it was explained in my junior high school biology course, around 1960. As an inquisitive ninth-grader I wondered,

"But what *is* a gene, and how does it determine those characteristics? What does a chromosome *look* like?" Neither my textbook nor my teacher offered a satisfying answer, and I made a note that this would be a fascinating question for me or someone else to research in the future. What I didn't know was that the question had already been answered, but so recently that the answer had not yet found its way into junior high classrooms.

NOW YOU SEE IT, NOW YOU DON'T

Mendel himself could not have known exactly what carried genetic information; the techniques for discovering the details would not be discovered until long after his death. What he could do was systematically describe a large body of observations and figure out a scheme by which it all made sense, and which made prediction possible. Reproduction involved unseen things (genes) that came in pairs. They could be dominant or recessive; and they determined characteristics according to whether corresponding genes from two parents were both dominant, one dominant and one recessive, or both recessive. This enabled Mendel to predict what types of offspring a given union could produce, and how likely each type was. That was a profoundly important first step.

Unfortunately, hardly anyone cared. Mendel was an amateur botanist with a lot of careful observations, but no theory to explain *why* they were as they were, and no generally recognized credentials. When he presented his findings to the local natural history society, they met with profound indifference. When he tried to get support from a prominent Swiss botanist named Karl Wilhelm von Nägeli, von Nägeli brushed him off. Mendel did publish, but in an obscure journal where he attracted little notice. Part of his problem may have been that Mendel's work itself was an unfamiliar kind of convergence, combining biology and mathematics in a way seldom if ever seen before, so that few potential readers were comfortable with both aspects of what he had done.

In any case, he grew weary of being ignored, he became preoccupied with administrative duties as abbot of his monastery, and he gained weight and became unable to handle the physical labor of gardening. His research slowed to a stop, and the few who had even heard of it forgot it. Mendel died, discouraged and never suspecting the regard in which his work would eventually be held, in 1884.

The story of Mendel and his findings reminds us that scientists are as human as anybody else, and offers examples of their best and worst behavior. I would not accuse von Nägeli of being evil, but it seems fair to say that he was at least a little arrogant and shortsighted. Few today have even heard of him. His most important effect on the history of science was probably that he delayed the development of genetics by failing to recognize the importance of Mendel's work and use his own prominence and influence to help it gain the attention it deserved.

On the other hand, toward the end of the nineteenth century three other botanists (one Dutch, one German, one Austrian), unknown to one another, independently did their own studies that essentially rediscovered what Mendel had found. Only afterward did they search the literature and rediscover Mendel himself—and not one of them tried to claim it as his own discovery. Instead, they brought Mendel's work back to light—more light than ever before—and announced their own as confirmation of his findings rather than a radically new discovery of their own.

One of them, though, Hugo Marie de Vries in the Netherlands, added something significant of his own. In his plant-breeding experiments he found that although usually the distribution of traits among offspring followed Mendel's rules, sometimes a plant would appear with a *new* characteristic, or mutation, which would then be passed on to its descendants. This observation was the key to overcoming a crucial weakness in Charles Darwin's original formulation of the theory of evolution.

As already noted, Darwin saw genetic variation as a gradual process, with characteristics drifting sometimes a bit this way and sometimes that way. But if that was the case, why would a new variant stay the same long enough for natural selection to work on it? It might just as well drift back in the next generation toward where it had started, or drift still further in the new direction. At a contrasting extreme, the German biologist August Weismann had made the observation that life itself (as distinct from individual organisms) seemed to be immortal, something he called *germ plasm* being passed from generation to generation in an unbroken chain of life. He even got the idea that only half of the germ plasm was passed on from each parent, the two halves combining to form a new whole in the offspring. To account for the immortality of germ plasm, he believed it must be protected somewhere deep within cells, and recognized no mechanism for variation to occur at all.

De Vries resolved the conundrum, and patched the weakness in Darwin's theory. He combined Weismann's germ plasm idea with Mendel's observations on the discrete nature of inherited characteristics—and his own observation of mutations. Since mutations were passed on without further change to subsequent generations, they must involve well-defined changes in the germ plasm itself.

But the question remained: What *was* the germ plasm? How did it carry information, what sorts of changes were involved in a mutation, and how were those changes inherited?

WHEN SCIENCES COLLIDE

The greatest revolution in biology began to take shape in the years around and following World War II, when other kinds of scientists began to take an interest in biological problems and apply their own tools to studying them. We have already looked, for example, at x-rays as a means of seeing inside living organisms—but they also had other, subtler applications at a submicroscopic level. In the applications we have already looked at, the incomplete absorption of x-rays by different kinds of matter was used to make recognizable pictures of the inside of the human body. In *x-ray diffraction*, first observed by German physicist Max von Laue and refined by English physicists William and Lawrence Bragg (the only father-son team ever jointly awarded a Nobel prize), the way in which x-rays are scattered or deflected by atoms is used to examine the submicroscopic structure of matter. The scattered rays interfere with one another, like crisscrossing ripples on a pond, and form patterns that don't look like pictures of anything we recognize. But they can be used to calculate details about how atoms are arranged, say, in a crystal—or in a giant molecule.

One giant molecule of very special significance is DNA (short for deoxyribonucleic acid). Back in the late nineteenth century, the German anatomist Walther Flemming, a pioneer in the use of dyes in microbiology, watched the process of cell division during reproduction. He noticed that objects within the cell—he called them chromosomes— doubled in number, then divided into two equal groups, each group going into one of the two new cells resulting from the split. He didn't know about Mendel, but when de Vries rediscovered Mendel's work, it became clear that chromosomes had to be where genetic information was carried, and their division and recombination neatly accounted for Mendel's observed rules.

Subsequent research showed that chromosomes were essentially bundles of DNA, so apparently the division and replication process acted on that molecule itself. Understanding *how* it operated required a detailed knowledge of how the DNA molecule was put together. That was a much bigger problem than it may sound like, and the traditional tools and methods of chemistry simply weren't up to the task. It took a convergence of several sciences to clear the hurdle, and the time was ripe for that to happen.

If you remember just a little high school chemistry, you may think of a chemical having a particular name as having a definite, unchanging formula and structure. Common salt, for example, is sodium chloride, represented by the formula NaCl, meaning it contains equal numbers of sodium and chlorine atoms. More precisely, in its solid form it consists of equal numbers of positively charged sodium ions (atoms that have lost an electron) and negative chloride ions (atoms that have an extra electron) that are arranged in a specific kind of crystal lattice.

DNA is a quite different kettle of fish. It's *huge*, containing many thousands of atoms, and not always in exactly the same proportions. So the things called *DNA*, unlike molecules of a simple compound like water or methane, are not all identical. However, early chemists managed to establish that all the things they were calling DNA were made up of smaller building blocks called *nucleoside phosphates*, each containing still smaller parts called *nitrogenous bases*. Those were of four types, adenine, cytosine, guanine, and thymine, and they seemed to come in pairs: adenine and thymine came in equal numbers, as did cytosine and guanine. But there was no predictable or consistent connection between the numbers of adenine-thymine and cytosine-guanine pairs. That observation suggested some connection between the paired bases, but what was it?

X-ray diffraction is where sciences began to converge on the DNA problem. Previously DNA had been seen as the province of chemists and biologists, but the first big break in the mystery came from a physicist, Maurice Wilkins (originally from New Zealand but working at King's College in England), who turned this new tool to a study of its structure. So did a colleague, Rosalind Franklin (English), who had her own research group at the same college, independently doing similar work. Analyzing x-ray diffraction patterns is complicated, and DNA in particular was so complicated that neither Wilkins's nor Franklin's results immediately gave a clear, detailed model for its structure. But they did suggest that it had some sort of spiral arrangement.

Franklin and Wilkins were not the only physicists getting interested in biochemical problems. Several of them converged on Cambridge University in England, where the Austrian-English biochemist Max Perutz had founded perhaps the first laboratory devoted to molecular biology— a blend of chemistry, biology, and physics that has since grown to be one of the most important of all fields of research. One of the physicists there was Francis Crick (English), who, with a young biochemist named James Watson (American), got interested in Wilkins's and Franklin's x-ray diffraction data on DNA. Trying to imagine a structure that would explain both the x-ray results and the way DNA replicated, they came up with the model now famous as the "double helix." (See plate 7.) In 1953 they published the paper announcing the model,[1] which was borne out by much subsequent research in many laboratories. In 1962 Wilkins, Watson, and Crick jointly received the Nobel Prize in medicine and physiology.

Watson published his own personal account of the work in his popular book *The Double Helix* in 1968,[2] but his view of it has since become in at least one respect highly controversial. He portrays Rosalind Franklin as subordinate to Maurice Wilkins, but in fact subsequent biographical research indicates that he had it backward.[3] A crucial step in inspiring Watson's and Crick's double helix model was an x-ray photograph that Franklin took and Wilkins showed to Watson and Crick without her knowledge or permission. She was very close to figuring it out on her own. In fact, she published an article of her own in the same journal issue as theirs, but its placement—and prevalent attitudes of the time toward women in science—encouraged the impression that theirs was the important discovery and hers mere corroboration. A very good case can now be made that she deserved to share the Nobel at least as much as Wilkins (some recent writers refer to Wilkins as "Franklin's colleague"), but since Nobels are not given posthumously (she died of cancer in 1958), the apparent wrong will never be fully righted.

Questions of priority and who deserves what credit aside, the essence of the double helix model and its relevance to genetics is this: the DNA molecule can be thought of as a very long, twisted ladder, in which each strand is a string of nucleoside phosphates and each rung is a pair of bases that fit together in a combination of the right length, either adenine (A) plus thymine (T) or cytosine (C) plus guanine (G). The bases can be arranged in any order along one strand, but their order along one strand automatically and completely determines their order along the other. If one strand contains the sequence CATGA, the corresponding section of the other strand has to be GTACT. The order of bases along one strand is the genetic code, consisting of four letters, in which the instructions for building any organism from a paramecium to a president are stored.

To reproduce, a cell divides. When that happens, the two strands of its DNA separate, rather like a zipper unzipping as the two bases of each rung break apart. Each of the resulting strands now has a row of bases sticking out, each of which would prefer to attach itself to another base of the kind that fits it. Metabolism is constantly making new nucleoside phosphates to use as building blocks, so there are plenty of those available to attach to the separated strands. When an unattached nucleoside phosphate containing, say, a guanine bumps into a cytosine sticking out from a separated DNA strand, it attaches itself. Wait long enough and a complementary block will attach itself to every protruding base—and once that happens, you have an exact copy of the original double helix.

So now we know what genes and chromosomes are. Chromosomes are long DNA molecules; genes are "words" consisting of particular sequences of base pairs along a particular part of such a molecule. Each gene determines, or helps to determine, a particular trait of the complete organism, such as dwarf or tall in Mendel's pea plants or eye color among your relatives.

Let us now look at how DNA provides the key to both reproduction and evolution. It carries complete instructions for making an organism. If replication were *always* exact, all organisms would be exactly like all their ancestors. Fortunately (for biological diversity, though often not for individuals) replication doesn't always work perfectly. Things like environmental chemicals or radiation can cause *mutations*, which are local changes in the DNA such as deleting a base or exchanging it with a different kind. Since such a change results in a different genetic word, something will be different about the way the resulting organism develops— for example, it may have a different eye color or too many legs. Usually those changes, being essentially random, are detrimental and the individuals experiencing them die early, so the change doesn't become established in the gene pool. Some don't kill their victims early enough for that to happen, but instead cause inheritable vulnerabilities to genetic diseases like cystic fibrosis or Huntington's chorea. But once in a great while, a mutation will improve an organism's ability to thrive in its environment, and that kind is passed on to many succeeding generations. If enough beneficial mutations accumulate, you can wind up with organisms so different from their ancestors as to constitute new species.

There's one more mechanism by which all but the simplest organisms improve their ability to try out new combinations of traits: sex. Instead of simply making copies of their own DNA, which usually means repetition of the same old same old, and depending on chance mutations to try anything new, sexually reproducing organisms combine a chromosome from one individual with a corresponding one from another. That will almost always lead to a combination of traits in the offspring different from either parent.

But the key to it all is still DNA, with its unique properties that were deciphered by bringing together the tricks of quite different scientific trades. And once scientists understood how the genetic code worked, they became very interested in exactly what it said—because if they knew exactly which DNA code sequences determined which characteristics of the organism as a whole, that would open up a huge range of possibilities—not just academic, but very practical and sometimes ethically challenging.

FROM DNA TO THE HUMAN GENOME

Knowing that chromosomes consist of DNA molecules with variable sequences of bases paired in a systematic way with counterparts on two intertwined strands is one thing. Finding out exactly what that sequence is in a fruit fly or a fly fisherman is a huge, difficult undertaking. But molecular biologists have developed ways to tackle it by breaking it (and the molecule under study) into small pieces, analyzing those, and then figuring out how to put them back together into the whole.

It's a daunting task, if only because DNA molecules are, by molecular standards, so big and complex, yet their components are so small by everyday human standards. The total package of DNA contained in a cell is called the organism's *genome*; it's the same in each of the organism's cells, and is divided among the chromosomes (of which there are a specific number for each species). The *smallest* known genome, belonging to a kind of bacterium, contains about six hundred thousand base pairs; that for a human (or a mouse, lest we get smug about our own complexity and presumed superiority) contains more than three billion.

Even though the problem is extraordinarily large, it's also of extraordinary interest to scientists, so molecular biology has grown like a snowball. Many workers in many laboratories developed pieces of the puzzle, and by the mid-1970s the tool kit for genome sequencing, or determining the order of the base pairs along the chromosomes, was assembled and in active use.

DNA sequencing is a process with many stages, all complicated, which I will sketch only briefly and qualitatively. (You can see more details, and even watch a video graphically showing the steps, at the Human Genome Project Information Web site.)[4] The first step is to break the chromosome being studied into much shorter, more manageable pieces. The key to this is called a *restriction enzyme*, a chemical that you can think of as molecular scissors for snipping a long chain into short fragments (a technique discovered by Hamilton Smith in 1969). Some restriction enzymes can be used to cut DNA at specifically chosen points, but the first stage of a typical sequencing doesn't bother with that.

Since even a single chromosome is so large, initially little is known about it, and the process is carried out in a solution containing many molecules. The DNA is initially sheared *randomly* into much shorter fragments (hence the name "whole-genome shotgun sequencing method"). Identical but different molecules in the solution will thus be cut in different places, so some of the pieces will have overlapping sequences.

Each of those fragments is used to make a new set of still smaller samples differing in length by one base pair, and those are separated by a technique called *gel electrophoresis* (a process that also depends on still another technology, that of dye making). Chemically identifying the bases at the ends of those subfragments makes it possible to piece together the base sequence in the piece from which the smaller pieces were made. This "reading" process is largely automated, using some of the same technology as computers. Each read gives a sequence of several hundred bases, and then computers are used to figure out (by matching up overlapping sequences) how the small pieces made in the first step had to have been put together in the original full-size DNA molecule.

Here again is a situation in which learning to understand one field— biology at its most fundamental level—is absolutely dependent on the convergence of other fields which at first glance might seem unrelated. Physics (especially x-ray diffraction) was needed to determine the overall structural plan of DNA. Advanced chemical techniques, including dye making and electrophoresis, were necessary to break DNA into manageable pieces, analyze those, and figure out how they must have been assembled. Computers—themselves a product of multiple convergences, as we have already seen—are indispensable for analyzing the results; as with medical CT scans, the sheer quantity of data and relationships could never be handled without them. And in the end, we come back to relating those results to the much earlier work of people like Mendel and de Vries to determine what it all means: which DNA sequences actually determine which characteristics of real organisms—and how.

The first genomes completely mapped were not from full-fledged organisms, but from viruses. Viruses stand in a sense between definitely alive and definitely not alive; they contain and reproduce by replicating DNA, but can't live on their own. Instead, they propagate by inserting themselves into the DNA of free-living organisms such as plants or animals. A virus of particular importance to human beings, HIV-1 (which causes AIDS), was mapped in 1985 by researchers at Chiron Corporation. The first free-living organism wasn't fully decoded until 1995, and that was a simple one: the microscopic, single-celled influenza microbe, analyzed by Craig Venter and his associates at The Institute for Genomic Research in Maryland. The first multicellar organism (a roundworm) was done three years later, and *Drosophila*, the fruit fly that had long been a favorite subject for genetics researchers, a year after that.

But even before those relatively modest aims were achieved, researchers had already set their sights on a far more ambitious goal. In 1988 the National Research Council announced its support for a national effort to map the entire human genome—yes, all the genes and all three-billion-plus base pairs of it. And the National Institutes of Health (NIH) set up an Office of Human Genome Research (HGP), headed by James Watson. Two years later the NIH and the Department of Energy (DOE) officially launched the Human Genome Project, a fifteen-year project to be carried out in laboratories all around the world—twenty of them, in six countries. (Why the DOE? It goes back to 1984, when a DOE conference considered using DNA research to trace mutations caused by the atomic bombs dropped on Hiroshima and Nagasaki in World War II.)

Meanwhile, Craig Venter had established his own company, Celera Genomics (a new kind of company, you will note, but by now a rapidly proliferating kind), and conducted his own private human genome project, in effect racing private against public enterprise to map the whole human genome.

It was, for practical purposes, a tie: both the HGP and Craig Venter's company had working drafts by 2000. The results were published the next year, in the journals *Nature* and *Science*. Meanwhile, all the work and computational power that went into getting those results had also spun off greatly improved sequencing techniques that were applied in other ways. Side benefits included genome maps for many other quite different species, and new insights into things like epidemiology and evolution.

After more refinement, the "final" report on the human genome was published in 2003—the fiftieth anniversary of the double helix.

And now the big questions are: Now that we have this vast wealth of unprecedented knowledge, what shall we do with it? What *should* we do with it?

CHAPTER 6
NEW DIRECTIONS IN BIOTECHNOLOGY

U p to this point, we have been looking at some of the major currents and convergences that have led to our present world, partly to understand where we are now and partly to gain some understanding of how the process works. We now begin to shift our emphasis, for if this book has a single central point, it is that our present world is by no means an endpoint or a finished product. Powerful currents of change are still at work and still converging, and new ones keep arising. From here on we will concentrate more on what major currents we can already see at work—and where they seem likely to take us in the relatively near future. Since we've just been looking at one of those currents, genomics, and since that and closely related fields promise to have some of the most profound impacts on how people live and think, we'll begin there.

The central importance of the genome is its crucial role in both reproduction and the development of individual bodies and minds. Those are also some of the most crucial and emotionally charged areas in both individual morality and social constraints. Societies, and the people who live in them, tend to have strongly held views about matters of sex, marriage, and the creation and rearing of children. When anything threatens (or promises) to overthrow or even suggest the possibility of modifying those views, some will eagerly (perhaps *too* eagerly) welcome the change, while others will vehemently (perhaps even violently) resist it.

The previous chapter only hinted at the potential for change in this area. In 1996, for example, much of the world (but not all of it) was shocked by the announcement that a group headed by Ian Wilmut at the Roslin Institute in Scotland had produced a sheep named Dolly, not by the usual collaboration between two adult sheep, but by cloning from a cell taken from an adult ewe. Several governments promptly responded with preemptive bans on cloning research, especially in humans, and today occasional letters to editors still surface decrying "the horrors of human cloning."

But exactly why is it horrible? For that matter, are we sure it *is* horrible? In most of those letters I see little evidence that the writers have given the question any thought at all. They usually just seem to be reacting reflexively to the fact that this kind of reproduction is different from what they were brought up to regard as "normal." Thirty years ago, in the late 1970s and early 1980s, we heard very similar outcries over the first human uses of in vitro fertilization (IVF)—the process wherein a human embryo is produced by combining egg and sperm in a laboratory and then implanting it in a womb to finish developing in the usual way. The first "test tube babies" were front-page material for sensational tabloids, and we still see occasional "Where are they now?" features about the first few. Yet the heated controversy has largely died down. IVF has become a widely accepted part of the tool kit for dealing with infertility. More than 1 percent of all births are conceived this way,[1] and more than one hundred thousand such babies have so far been born in the United States alone.

Is it possible that cloning will in time achieve a similar level of acceptance? If so, is that a good, bad, or indifferent thing?

Those are questions of a kind we shall increasingly have to confront and deal with. Simply ignoring them and hoping they will go away is not an option. Once people *can* do something, some of them will want to— and either they will, or others will have to find ways to prevent them (and with better reasons than "I don't like it").

SOMETHING NEW, OR AN EXTENSION OF THE OLD?

The word *biotechnology* is relatively new, as are many of its forms. But the basic idea of using living things to do or make things for us, and reshaping them to better serve our wants, has been around for a very long time. As Jared Diamond discusses in his book *Guns, Germs, and Steel*, the origins and growth of technological civilizations on Earth depend ultimately on the fact that a few species of plants and animals lent themselves to domestication as convenient sources of food, drink, manufacturing materials, and usable power.[2] That in itself is an early form of biotechnology: corn and cotton plants are organic "machines" used to make food and fabric; cattle and goats for making food, milk, leather, and glue.

But the original wild forms of corn, cotton, and cattle were not nearly as good at those processes as the ones we now take for granted. Modern varieties are the result of the first sort of genetic engineering: not tinkering directly with DNA, but empirically finding new ways to combine it by choosing which plants or animals to use as breeding stock, and in what combinations. Using Mendel's laws long before Mendel formally codified them, farmers and ranchers have long bred new varieties of crops and livestock for specific useful traits like flavor, drought resistance, high yield, strength, docility, or intelligence.

Not surprisingly, people have tended to regard themselves as a special case, to be treated with a complicated mixture of emotional and intellectual considerations. Even as they made practical use of selective breeding, and easily recognized some breeds of dogs as characteristically having different temperaments from others, they have alternatively embraced and reviled the idea that the same principles might apply to us. At some times and places "everybody knew" that some "races" were intrinsically brighter or more industrious than others. At others (like here and now), it is considered offensive to suggest that statistical differences exist between one group and another. As for deliberately trying to breed humans for health or intelligence, we've often been ambivalent. Eugenics got a bad name, for good reason, when Hitler and his henchmen were ranting about creating a pure Aryan race of "supermen." Yet virtually all prospective parents base their choices of partners, consciously or unconsciously, at least in part on trying to maximize their chances of having bright, healthy, attractive children.

The new era in biotechnology arguably begins with the understanding of DNA gained in the middle of the twentieth century (as discussed in the previous chapter). Knowing where genes were, what they did, and how to manipulate them, researchers soon learned to take DNA apart and put the pieces back together in new ways, creating *recombinant DNA*. Such DNA, having a different base sequence and gene sequence than the original molecule or molecules from which it was derived, would, if allowed to replicate, lead to the growth of a new and different organism.

On casual hearing, this may sound a lot like what happens all the time in sexual reproduction, but this can go way beyond that. In sexual reproduction, you start with two DNA molecules from the same kind of organism, each containing a particular sequence of genes coding for a particular set of characteristics. Each may contain a different version of a particular gene—for example, one for brown eyes and one for blue—but they both contain specifications for the same set of variables. So when you combine one from one parent with its counterpart from the other, the result is still clearly a chromosome pair for the same species, and coding for the same set of characteristics.

In recombinant DNA, the recombination may instead occur by snipping whole pieces of a DNA chain and putting them back together in a different order. It's even possible to put together pieces taken from completely different organisms—even organisms so different that we wouldn't normally think of them as even distantly related. In 1973, to give one of the earliest examples, Stanley Cohen and Herbert Boyer introduced frog DNA into the *Escherichia coli* bacterium, which then reproduced with the combined DNA. In 1980 a group headed by Martin Cline created the first transgenic mouse. Like much basic research, those early experiments were more for learning the principles and techniques than for practical application. But also in 1980, a human gene responsible for making the protein called *interferon* was introduced into, and reproduced by, a bacterium—which could help patients who can't produce adequate interferon for themselves.

Combining human and nonhuman DNA makes many people nervous,

even when it has such clear benefits as providing an efficient way to make a lifesaving drug (such as insulin, the gene for which was cloned in 1978). In 1978, 150 molecular biologists, anticipating the potential dangers, held a conference about the need for controlling genetic research, which led to strict government regulation.

The government got involved in another way in 1980, when Ananda Chakrabarty genetically engineered a bacterium that could break down crude oil and tried to patent it. The US Patent Office had previously held that living things could not be patented, but this case was so unprecedented that it went to the Supreme Court, which ruled (narrowly) that artificially developed organisms, unlike newly discovered natural ones, *could* be patented.

That opened the gates for a veritable flood of genetic engineering research and development. One company dedicated to the field, Genentech, had been founded as early as 1976 by Herbert Boyer and Robert Swanson, and others soon proliferated. In 1982 the Food and Drug Administration (FDA) gave its approval to bacterially produced human insulin, the first recombinant DNA product approved for medical use. In 1986 the FDA approved the first genetically engineered vaccine, for preventing hepatitis B.[3]

Medicine is not the only use for genetic information and engineering. Crops have been "gengineered" for traits like ability to resist diseases, thrive in new climates, or stay fresh longer. Predictably, many people are uncomfortable with that, too; you've undoubtedly heard of customers refusing to buy, or vendors refusing to sell, "Frankenfoods"—usually with little or no consideration of whether they're actually better, worse, or not significantly different from their unmodified counterpart.

Since DNA essentially defines an individual, it provides an even more positive means of identification than fingerprints. DNA fingerprinting, having no necessary connection with fingers, can be done using just about any tiny sample of a person's body, such as blood, skin, or hair. Since tiny fragments of such things are commonly left behind wherever a person goes, DNA identification has become a standard and powerful tool for criminal investigations and paternity suits. (Though acceptance did not come instantly or unanimously:[4] my local newspaper recently featured extensive coverage of a man who was recently released from a life sentence and exonerated when somebody else confessed to a murder for which he was convicted sixteen years earlier—even though DNA evidence at the time showed he couldn't be guilty.)

Since individual genes are linked to susceptibilities to particular diseases, DNA analysis can be used to screen for such vulnerabilities. It can identify a person as likely to develop, say, heart disease or Huntington's chorea, long before he or she actually does so. Furthermore, not all individuals react the same way to the same substance: the same drug may be a lifesaving cure for one person, and a deadly poison to another. Genetic engineering opens up the possibility of recognizing those tendencies and custom-tailoring a drug to a specific patient, as demonstrated in a burgeoning new field called *pharmacogenomics*.[5] Similarly, genetic differences cause some patients to have difficulty metabolizing certain foods, or even to have life-threatening allergies to them. Learning to identify those cases and tailor diets to individuals based on their genetic makeup has triggered the growth of another new field called *nutrigenomics*.[6]

I've already mentioned Dolly, the first full-fledged mammalian clone. Horticulturalists have long propagated some plants by cloning, and some so-called lower animals had been cloned (some species of lizards reproduce only that way, consisting entirely of identical females). But a sheep is just enough like a person to strike closer to home and raise alarms among people who had never thought of cloning as having any real relevance to them. With Dolly, cloning just about anything—including people—suddenly seemed a real, and perhaps imminent and frightening, possibility.

And if you put the possibility of cloning together with those opened up by recombinant DNA, horizons widen dramatically. It becomes possible to create new kinds of organisms by combining traits of unrelated species—for example, the genes responsible for light production in fireflies can be introduced into aquarium fish to create luminous fish for the pet trade. People wonder how far such combinations might go, especially if some of the genes involved are human. What if someone wanted to, and could, produce a person with a venomous bite, or a wolf with the intelligence of a person?

Such notions have long been the stuff of bad science fiction, and no doubt some of them are unrealistic and will never go further than that. But it's already clear that the range of real possibilities is far more extensive than almost anyone would have assumed just a few years ago, and that we are just beginning to glimpse their possible extent.

Of course, cloning is not as simple as it may sound in the casual descriptions above. In pre-Dolly days, many considered advanced animal cloning—growing a complete adult animal from a single adult cell—unattainable because adult cells seemed to lose the ability to produce other types of cells. Remember that ordinary reproduction starts with a single egg cell and a single sperm, containing a full set of "blueprints" and the ability to produce just the right number and arrangement of all the specialized cells needed by a complete organism: bone, skin, muscle, nerve, heart, liver, and so on. Cells differentiate into all those types as the organism develops, and their DNA acquires imprinting that "switches off" some of the genes and makes it impossible for liver cells, for example, to produce pancreas or brain cells.[7]

It turned out, though, that the imprinting patterns of "stem cells" in early-stage embryos are similar enough to the original fertilized egg to still have the ability to differentiate into all types of cells. This was demonstrated on mice in the 1980s, and around 2000 for human beings.

The process of IVF involves producing multiple embryos in the laboratory, freezing some of them for storage, and selecting small numbers of them to implant for development into babies and ultimately adults. Many of them would be discarded or kept in storage indefinitely; in either case, they would never develop into complete, sentient human beings. Might it not be better, and morally defensible, as many have argued, to harvest some of those and use their stem cells for purposes such as growing replacement organs for people currently far down on waiting lists for transplants?

"Morally defensible" depends on whom you ask, and the question remains heatedly controversial. There are also practical problems, such as the fact that organs grown in this way might be rejected by a recipient—unless, of course, the recipient was the same as the donor. That sort of thing is already done with blood transfusions. *Autologous transfusion* is the practice of banking one's own blood for transfusions before an anticipated operation, to be sure the blood you get is compatible and uncontaminated.

Doing the same thing with other organs might seem impractical, if it requires embryonic stem cells. But we can already see at least two possible ways around that problem. First, not long after the potentials of embryonic stem cells burst onto the scene, researchers found that adult stem cells, undifferentiated cells with some of the same capabilities as embryonic cells, still existed here and there in adults. Those, they thought, might provide a way to do "therapeutic cloning" such as growing replacement organs for a specific patient, while bypassing the moral dilemmas and controversy surrounding the destruction of otherwise unused embryos for their stem cells. Second, while it's too late for people who are already grown up to harvest their own embryonic stem cells for future therapeutic use, it's now possible for parents to plan at birth to provide that option for their offspring. It has already become fairly common for new parents to have their newborns' cord blood, taken from the routinely discarded umbilical cord, collected, and stored as a future source of such cells.

As for full-fledged cloning of viable humans and other mammals, that possibility is now well established and in some (nonhuman) cases already in use. Sometimes, as in agriculture, cloning can have definite advantages, though those must still be weighed against potential customers' emotional objections. A few beef producers have cloned cattle; some customers have sworn emphatically that they would never eat such stuff. To make sure they don't do so unwittingly, some have demanded legislation requiring cloned meats to carry warning labels. In the absence of such legislation, some vendors have jumped on the bandwagon and publicly promised that they would not sell cloned meats.

These attitudes may well shift in the future. No study I've heard of has found any difference in nutritional quality or safety between cloned and conventionally raised meats. In fairness to those who are uncomfortable with the idea, I must acknowledge that the success rate of early mammalian cloning efforts was not very high, and that an unusual number of early cloned mammals died prematurely. (This was apparently due to imprinting problems having to do with the conditions under which the cells developed, not defects in the DNA itself.) It is too early to conclude from those facts that cloning is inherently unreliable or unsafe. Any new technology starts off crudely and needs refinement. Early automobiles were primitive, cranky (literally and figuratively), and unreliable—a far cry from today's versions. Quite likely the problems experienced with early cloning efforts are just that: problems that can and will be solved.

On the other hand, that doesn't mean we should start doing it whenever and wherever we can. Technical problems are only one of the kinds we face; the others are likely to prove even harder.

WHAT WE *CAN* DO VS. WHAT WE *SHOULD* DO

The discoveries already made in biotechnology, and the ones we can anticipate "downstream," open a great many possibilities that could radically change the way we live—and may require some rethinking of moral, ethical, and legal codes because the old ones don't really cover the new choices we'll have to make.

For example, should genetically engineered or cloned foods be banned or require special labeling or governmental regulation? If genetically engineered or tailored drugs lead to dramatic increases in life expectancy, will that exacerbate the problems, like global warming and overfishing, already being caused by a large and still rapidly growing population? If so, what can be done to minimize or reverse the damage?

If population growth is indeed causing sustainability problems that may become catastrophic within this century, as many scientists believe,[8] should we even be trying to prolong lives or increase fertility by every means possible? Almost everyone would agree that those are good things, when considered only at the level of individuals directly affected by, say, the ability to have a child or to live an extra five, ten, or fifty years. But if everybody does those things and nothing else changes, population *will* grow even faster—and, along with it, so will things like resource depletion and the need for more governmental control.

Similarly, being able to feed more people is certainly one of the most desirable of goals, if you're one of the hungry. But one of the overarching themes in the history of civilization is an ever-spiraling cycle of cultures becoming able to feed more people, needing more people to do the work of feeding them, expanding the population still more because they can, and so on. How far can that go? Genetically engineered crops can surely fuel more rounds of this—but should it? Or at some point do people need to find a way to say, "We're going to do this to make life better for all of us, but we're going to slow down the rate at which we make more of us?" If so, how will that happen, given the powerful influence of cultural and religious attitudes toward such matters?

IVF not only allows couples who otherwise couldn't to become parents of their own children, but also opens up a whole complex of other options about which traditional legal and moral codes provide at best vague guidance, since nobody anticipated them when those codes were shaped. Some of those issues courts and churches and individuals have already had to confront; others haven't arisen in the real world yet, but surely will.

The womb into which an in-vitro-fertilized embryo is implanted is not necessarily the one from which the egg was originally taken. What are the roles and rights of egg and sperm donors and surrogate mothers, and how are those reconciled with the roles of "natural" parents? There's some precedent for all this in the roles of birth parents and adoptive parents, but there are also many new and, for many, uncomfortable questions.

Parents now have at least two ways by which they can choose the sex of their offspring: either by using only selected sperm in IVF, or by bringing only embryos of the desired sex to term. There are many reasons why parents might want to do this, ranging from a simple desire for a "balanced" family, to a general cultural bias viewing one sex or the other as more valuable. But should they—and should they be allowed to? Most existing societies have woven deeply into their fabric a wide range of customs and institutions predicated on the assumption that the population will include very nearly equal numbers of men and women. Usually, with nature left to its own devices, this will be true. What if it becomes fashionable for most people to choose the sex of their children, and for a few years a majority of them are choosing girls (or boys)? A few years later, society will have an unusually large number of women (or men) who can't find partners. Is this necessarily bad, and should it be allowed to happen? If so, should it be accommodated by changes in customs and laws?

Once children are born, our already impressive and rapidly improving knowledge of the relationship between DNA and what happens to the body is beginning to allow things such as identifying people at high risk of developing brain cancer or cystic fibrosis. Would you want to know now that you're likely (but not certain) to develop a deadly disease in ten, twenty, or thirty years? Or would you rather wait and see, not knowing whether your risk is higher than anyone else's? It's not a simple question. Not everyone would answer it the same way—and how anyone answers is likely to be very different if (a) something can be done now to eliminate the danger, or (b) the disease is neither preventable nor curable.

And it may not be a purely individual decision. Once the capability exists to identify individuals at high risk, might they be denied employment or insurance for that reason?

The same knowledge of how to read and manipulate DNA that makes potential disease identification possible also suggests a way to prevent that particular danger. Some genetic problems can be corrected by gene therapy, in which DNA with a defect repaired is introduced into a body to replace the flawed version. But what if it's done *before* birth? Embryos used in IVF, or even those fertilized the old-fashioned way in their mothers' bodies, can be screened early in their development, and genetic problems repaired then so they won't cause trouble later.

But who decides what constitutes a "defect" that should be "fixed," and what is just a trait that helps define an individual as who he or she is? If it becomes easy enough to do this sort of thing, we can imagine it being done routinely as part of standard prenatal care. And if that's done, we can just as easily imagine babies being homogenized to state specifications, or parents ordering "designer babies" and vying for status on the basis of what features they can afford.[9]

Or people might want to clone themselves—or a loved one they've lost.

The possibilities can be a bit dizzying, even overwhelming. That doesn't necessarily mean we should automatically shy away from them, but it does mean we need to think carefully about how, if at all, we use them. With all those new possibilities come new responsibilities and dilemmas. How can we prepare ourselves to deal with the choices?

A VACCINE FOR FUTURE SHOCK

When I first mentioned Dolly the lamb, I said that much of the world, but not all of it, was shocked by the announcement. Who *wasn't* shocked?

One group that generally found Dolly's cloning unsurprising, and not necessarily frightening, was science fiction readers. Science fiction writers had been talking about cloning as a real possibility for decades; regular readers knew it was only a matter of time till it happened. So when a reporter from a national news magazine called me at the office where I edit a science fiction magazine to ask how I felt about Dolly, my first inclination was to answer, "Ha! We told you so."

If you're not very familiar with science fiction, your first inclination may be to say, "So what if science fiction writers were writing about cloning? That was no reason to think it might ever become reality." If all you know about science fiction is what you see called by that name in movies and television, you might even have some justification for that view. Most science fiction in the visual media has tended to aim for little more than sensationalism and special effects, with little or no attempt to make its action plausible or believable. The "science" in the label has little or nothing to do with real science.

But much *printed* science fiction, published in magazines and books, has been a very different thing. Thanks largely to the influence of an editor named John W. Campbell who was trained as a physicist and took over the magazine *Astounding Science Fiction* in 1938, many writers have made a serious effort to keep up on developments in real science and imagine how they might affect human life in the future. Each story can be thought of as a "thought experiment," an attempt to try out a possible future and think through what it might be like before we have to deal with it in reality. A great many carefully researched and written stories had already dealt in this way with cloning and many of the other possibilities raised in this chapter, beginning long before Dolly shocked the unprepared. (See, for example, Theodore Sturgeon's "When You Care, When You Love,"[10] Ursula K. Le Guin's "Nine Lives,"[11] and Michael F. Flynn's "The Adventure of the Laughing Clone,"[12] or any of numerous other stories mentioned in *The Science Fiction Encyclopedia*'s article on clones.)[13] In a real sense, the best science fiction can act as a "vaccine against future shock."

Of course, not all science fiction is created equal. (Theodore Sturgeon once said, "Ninety percent of science fiction is crud—but then, ninety percent of *everything* is crud.") Some stories are still published that aren't careful with their scientific underpinnings and extrapolations, and a far larger number are written but not published. As an editor, I see a lot of those, and in ways, the things I often see in rejected stories may be more telling than anything published about popular attitudes toward possible future developments.

For example, I still see far too many stories in which the plot hinges on whether somebody is a "real person" or a clone, or whether one clone's killing another is murder or suicide. A clone *is* a real person, genetically related to his or her prototype in exactly the same way as an identical twin. (A writer friend of mine who happens to be an identical twin once began an essay with the words, "I am a clone . . .") And a clone is *not* the prototype, but a separate person, with the same genome but shaped by a whole lifetime of unique experience—so if one kills the other, it is clearly murder, not suicide.[14]

People may (and will) argue about whether people should be created by cloning; but once they exist, there will be no logical justification for treating them any differently than the ordinary rules of morality and courtesy require that *any* human being be treated by other human beings. Of course, there will be situations, notably in the legal arena, in which the old rules *can't* be applied in the accustomed ways, or where it's not clear how they should be. What happens, for instance, if a man dies without a will, and he's married but has also, for whatever reason, cloned himself? To what share of the estate, if any, is the clone entitled?

What happens if (as in Michael F. Flynn's story cited above) DNA clearly identifies a murderer, but it then turns out that the suspect has five clones, all with the same genome? In such a case, DNA fingerprinting is only the first step. It proves only that one of the six is guilty, but other evidence will be needed to identify and convict the only one who *is* guilty—while the other five are innocent.

The prevalence of those "real vs. clone" stories suggests a still more disturbing possibility. Even though no logical case can be made for treating a cloned person as less human than any other kind, or any less than a full and unique individual, the fact that so many people write stories in which they are treated that way suggests that in the real world, they might be. Neither laws nor popular attitudes are always logical or reasonable, and it's all too easy to imagine new classes of people being dehumanized and mistreated for reasons that make no sense.

People who want to avoid doing that will need to think in advance about the fact the future can and almost certainly will include new kinds of human relationships. We've already seen (and to some extent, adjusted to) test-tube babies, surrogate mothers, and sex changes. If we want to minimize unnecessary frictions, prejudices, hostilities, and outright atrocities in the future, we will need to think *before* the new possibilities materialize about how we should and will deal with them.

In other words, more of us will need to begin thinking like the best science fiction writers—and a good way to start might be by reading some of the best science fiction that has already tried to examine those questions.

CHAPTER 7
COGNITIVE SCIENCE
How Do We Know?

D NA is only one of the built-in information systems of huge significance to human life. Another is our nervous system: the brain and
all the apparatus with which it takes in and processes information about
the world around us, and enables us to do things in response. Since the
technologies we use enable us to extend and enhance our built-in abilities, using them involves interactions between external and internal
machinery. To get the most out of anything else we do, we must try to
understand how our minds and nervous systems work, so we can interact
as smoothly as possible with the tools we use.

The subject matter of psychology is subtler and more complex, and
more controversial among its practitioners, than that of the physical sciences, such as physics and chemistry. We are much more emotionally
invested in wanting to believe certain things about it whether they're true
or not. So psychology is not as precise and quantitative a field as electrical or aeronautical engineering (though we can't rule out the possibility
that someday it may be). But we can learn and apply some things about
it through observation and experiment. The particular branch of psychology dealing with how we perceive and learn and how we know—

cognitive science—has recently attracted much interest as one of the keys to bringing technologies together to make them as helpful as possible to their human users.

It has become a cliché in some circles to say, "The human brain isn't a computer." But in fact, it is: it processes information both from external inputs and stored memories; comes up with new results; and produces outputs based on those results, ranging from swinging a hammer in the right way to drive a nail, to filling out an income tax return, to writing or performing an opera. But the brain is a very different *kind* of computer from the manufactured tools we usually call by that name.

The computers many of us now have on our desks are extremely efficient at some kinds of computation. They are far faster than human brains at number crunching. Since other kinds of information can be encoded digitally in numerical form, they are also far faster at many tasks that we don't normally think of as numerical, such as finding and organizing data in huge files for a library catalog, changing the name Henry to Alistair every time it's mentioned in a novel, or transposing an entire musical composition into a different key. It's also far more reliable than most human brains at those sorts of things. A human proofreader may overlook a Henry or two, but the computer will scrutinize every word in the book and catch every one of them, with a very high degree of certainty. When you tell it to remember something, it does so, immediately, completely, and exactly. Component malfunctions can occur, but these days they're very, very rare, and good computer programs have checks built in to notice and correct for them.

On the other hand, information is stored in very specific places, and if the part of a disk where a chunk of data is stored is damaged, the data is lost completely and irretrievably. (Hard disk recovery programs that retrieve data can't do that if the data is truly gone; usually what happens is that the data is still there but the road map for finding it is damaged.) Computers (at least so far) are very literal-minded. They do *exactly* what they're told to do when particular conditions arise, but they have little capacity for taking initiative or making imaginative leaps.

Our built-in brains are, for some kinds of tasks, much slower, but they have advantages because of qualitative differences in how they work.

Instead of a simple electrical current flowing in a wire, nerve impulses involve chemicals called *neurotransmitters* that jump from one neuron (nerve cell) to another across a gap called a *synapse*. Data is not stored or transmitted in such a straightforward digital code; it's not even clear that *digital* is really an appropriate term in reference to the nervous system. The strength of a sensation depends, for example, on how frequently impulses are sent across a synapse.

It's hard to pinpoint the exact location where a memory is stored in the brain. There's some evidence that memories are "distributed," or stored partially in multiple places. This has the beneficial results that even if one spot in the brain is damaged, some of the memories stored there may still be retrievable elsewhere, and one part of the brain can take over jobs formerly done by another part. When we try to remember something, it may take numerous repetitions and the creation of mnemonic aids to make it stick. Our minds may wander and we may occasionally overlook or forget things. But we excel at putting together seemingly unrelated bits of information, and coming up with new and useful conclusions that we could never have reached by following simple logical paths.

How do we do those things? There's a great deal that we still don't understand about the detailed mechanisms. But it's clear that we can gain a lot by learning to let the special strengths of each kind of computer complement each other in as many ways as possible. We can see some good examples of where this might lead by looking at what's already been accomplished in trying to enable humans and computers to interact more easily, comfortably, and productively.

MAN/WOMAN TO COMPUTER: LET'S TALK

When I first started using computers, in the late 1960s, they were big, intimidating, relatively stupid monstrosities that my fellow students and I approached with trepidation.[1] A single computer was *the* computer for the faculty and students of an entire campus—and this was a university known as a major center for science and technology. The computer, with its peripherals (input, output, and storage devices), filled most of a large room, even though its memory and computing capacities were much smaller than even a cheap modern computer. (You can't even find such a device for sale now.) Since it was so slow and had such limited resources, anyone who wanted to use it had to reserve and pay for time. To get the most out of that time and money, we had to estimate carefully (which wasn't always easy) how much time we needed. In case we guessed wrong, we had to specify a time limit for each run, so that if the computer hadn't reached a conclusion in the time allotted, it would stop instead of continuing to run up the bill.

When the time came to do the run, we would gather up our programs and data and traipse across campus to the "temple" to make our "sacrifices." Hardly any software was available off the shelf, so unless you knew somebody who had written a program for a problem like yours, you'd probably have to write your own. Typically those were very specialized,[2] rather than versatile and widely applicable like the word processors and spreadsheets we now take for granted. The programs were written in specialized languages like Fortran and Cobol, bearing a slight, superficial resemblance to English, but full of cryptic statements like $I = I+1$ and READ (11, 47) X(IBOD), Y(IBOD), Z(IBOD). To load a program and the data to be processed, we had to punch each program command or group of numbers into a card, in a rigidly specified format, using a machine vaguely resembling an exceptionally clunky typewriter. Doing this wasn't hard, but it was so tedious and time-consuming that researchers who could afford it hired key-punch operators just to do it.

So, with armloads of punch cards held together with rubber bands or stacked in boxes, we arrived at the computer room and waited our turn. When it came, we'd load our stacks of cards into a stove-sized card reader, punch a button, and watch anxiously as the machine, like a mechanized blackjack dealer with an oversized deck, whisked cards off the loaded stack, read them (remember the Jacquard loom?), and spat them out into another stack. Other waiting users stood around watching, partly cheering us on and partly (secretly) hoping the machine would chew our cards up or spit them out in a fountain of unruly cardboard (as it sometimes did theirs). Our colleagues were still gathered around when the printer (another sizable machine that tractor-fed long rolls of big, perforated paper and had exactly one font) began spewing out results. On a good day, those were neat tables of never-before-known information that would have taken us weeks or months to calculate without the computer (which is why we put up with it). On a bad day (and there were plenty, especially when we were trying out a new program), they might just say something like FATAL ERROR, followed by a cryptic explanation of why our program had failed to do what we'd asked it to.

Needless to say, the phrase *user-friendly* hadn't been invented yet.

When the first personal computers came along, things had improved somewhat. It was no longer necessary to load all the data in a batch and then wait helplessly for final results. Data could be input, interactively, through a keyboard sitting right in front of a CRT monitor which showed us what we were working on. I could work on my project with some semblance of privacy while someone else worked on hers elsewhere in the same room. But telling the computer what to do with the data we put in still required keying in arcane commands that had to be memorized or looked up. Many people continued to regard computers as mysterious, intimidating, and beyond their abilities; and for some years, they remained an eccentric plaything of geeks interested in computers per se or as a way to play games.

One of the big factors that changed that, of course, was the development of the internet, which caught on as a new way of corresponding and shopping. But another crucial influence was the development of graphical user interfaces (GUI), which, for most people, felt far more natural and intuitive than typing long strings of seemingly nonsensical characters.[3] This was the central innovation in Apple Computer's Macintosh, introduced in 1984 as "the computer for the rest of us." Ordinary words and numbers were still typed on a keyboard, but a great many other processes were reduced to simple matters of looking for what you wanted to do and pointing at it.

Application programs and documents were represented by icons, small pictures on the monitor screen. Those were organized in nested folders, represented by icons looking just like file folders on the desktop of the monitor screen. To open a folder you would simply point at it with a "mouse" and press the click button twice. The screen would show that folder opening up, with its contents now spread out on the desktop. Double clicking one of those would open it in the same way, showing the document as it would appear on paper. To work on it, you'd use the mouse to select the part of interest, then choose what you wanted to do with it from a menu pulled down from the top of the screen.

Making this kind of interface was clearly a triumph of computer engineering, but what made it worth doing was a qualitative understanding of the kinds of things people find easy and difficult. Older interfaces required users to learn or look up long lines of coded commands, which few people found easy or comfortable to do. Those commands would be different for every application (such as word processing or database management). The Macintosh GUI required users only to look for what they wanted to do, either as a picture or as a word or two in a list, and point at it. Everybody is used to that type of organization in everyday life: practically all of us have ordered from restaurant menus, or stored papers in file folders so we could find them easily later. Furthermore, Apple used those same basic methods in all kinds of applications, so once you learned one application, it was relatively easy to learn another.

By taking advantage of knowledge about how people think and learn, Apple had succeeded for the first time in making a computer that could be used with relative ease by almost anyone, even people lacking the interest, patience, and/or aptitude to deal with long lists of decidedly non-intuitive commands. I vividly remember the enthusiastic words with which a writer friend who had just acquired one of the first Macs described it: "User friendly? This thing's a user *slut!*"

You might think that a system with such clear and dramatic advantages would take the world by storm, and quickly crowd out its older, clunkier competitors. It did, but not quite as quickly or straightforwardly as you might imagine. Apple had built a better mousetrap (no pun intended), but it viewed its hardware and software as a package deal. If you wanted to use the Macintosh graphical interface, you had to buy a Macintosh computer—and they were significantly more expensive than comparable models from other manufacturers. But it wasn't long before Microsoft developed its own operating system with a graphical interface, called Windows, widely adopted by IBM and manufacturers of "PC clones." It looked and felt a lot like the Macintosh system—enough so to cause some spirited courtroom discussion. When the dust had settled, both Macintosh and Windows remained on the market—and they still do (though Windows still has a much larger market share).

In a field that has long been characterized by dizzyingly fast change, this is a remarkable achievement. Computer hardware evolves so fast that many people consider machines only two or three years old as obsolete. (Moore's Law states that computer power doubles every eighteen months.) Yet the basic form of the user interface, with icons, mice, and menus, has become pretty well standardized across manufacturers, and has remained so for a couple of decades—because it works well with people, not as early computer manufacturers might have liked them to be, but as they are. It has given computing a revolutionary impact not only on how private individuals work, play, and interact, but also on how all kinds of business are done. It has radically transformed the conduct of publishing and scientific research.

And yet, it does have its limits. This more or less "standard" graphical interface is very visually oriented, and while that works very well for most people, it doesn't for all. What about those who are blind or visually impaired? An obvious direction for future development in fitting computers to people is to take into account that not all people have the same sensory equipment or learning styles. During my years of classroom teaching in a small college where I got to know all my students individually, I never met any two whose minds worked quite the same way. As our future becomes increasingly technological, we will need to find ways to make all our devices at least as accessible to all kinds of people as the graphical interface has already made computers to most of us. Ideally, we should be able to converse with our machines in whatever form of our own natural language we find most comfortable, be it spoken English or written Turkish. Small steps have been made in that direction, but it's a long and challenging road.

Meanwhile, what we have already learned about how people learn and interact with their machines has led to improvements in a variety of areas, notably education. Many schools have integrated personal computers into courses in all areas. "Smart blackboards" are beginning to take an active role in many classrooms, and interactive instructional programs (such as the Rosetta Stone language courses) are now available for those who don't have access to classrooms or simply prefer to learn on their own.

All of this has happened without detailed knowledge of exactly how brains work. We could see that a graphical interface would be a good idea from empirical observations about how people learn best, and what sorts of things they find easy or hard. How much more could we do if we understood *how* we think and learn, in the kind of detail that computer engineers understand computers?

WHAT'S GOING ON IN THERE?

People who design computers and software to run on them cannot think in such vague terms as simply describing what sort of input leads to what sort of output. They have to know, step by step, how one leads to the other: exactly what changes occur in all the circuits and memory devices along the way. This is necessary because they have to build the devices that will make it happen.

The brain, on the other hand, is a finished product. The best we can hope to do is "reverse-engineer" it—to figure out what happens inside it to produce the results that we observe. Our efforts to do this are limited not only by the complexity of the system, but also by ethical constraints on dissecting our fellow beings while they're in good working order. But I've already hinted at some noninvasive approaches that can already be used to get at least a coarse picture of what parts of the brain are actively involved in particular kinds of mental activity: positron emission tomography (PET) and functional magnetic resonance imaging (fMRI). There are also older ones such as electroencephalography (EEG), which maps electrical activity in the brain.

In general, those techniques still don't give a fully detailed picture of what's happening at the most basic, neuron-by-neuron level; but they are a significant step in the right direction. And sometimes they do give fairly detailed results.

In the early 1990s, for example, a group of neuroscientists at the University of Parma in Italy, while studying motor systems (the parts of the nervous system that control bodily motions) in monkeys and humans, discovered a group of neurons with an unexpected and peculiar, but apparently very important property.[4] PET and fMRI scans showed that certain groups of neurons "lit up," indicating activity, when their owners did specific things like lifting a cup. A smaller group of those neurons lit up in the same way while watching *someone else* do the same action!

Furthermore, they lit up in different ways depending on the intent behind the action—for example, whether the cup was being lifted to take a drink or to take it away for cleaning. In short, those neurons seem to reflect in an observer the actions and even the meaning behind the actions being taken by the person (or monkey) being observed. For this reason, they have come to be called *mirror neurons*, and they have become the object of vigorous research in several laboratories around the world.

The interest in mirror neurons is not mere academic curiosity. They seem to play a central role in such essential human functions as learning to do something by watching and imitating someone else doing it, as well as learning language, reading the emotional state of another, and feeling empathy—all of which are deeply embedded in the foundations of culture. When mirror neurons don't function normally, they can cause personality disorders that make an individual unable to interact normally with other people and their environment. Autism, in particular, seems to have a very strong correlation with nonfunctioning mirror neurons. They may turn out to be its principal underlying cause, and understanding of how the process works—or doesn't work—may lead to better ways than we now have of helping autistic children.

This is but one example of the kind of benefit that can be gained by looking at correlations between particular behaviors and particular kinds of brain activity. Others are likely to lead to still more understanding of how our minds function and interact, which may lead in turn both to new ways of dealing with mental illness and to new ways of helping everyone learn, work, and generally live to his or her maximum potential.

FROM ONE MIND TO MANY: THE MEME CONCEPT

Much of the special importance of mirror neurons lies in their key role in enabling humans to understand and empathize with one another. To optimize life for human beings, it is not enough to understand how each of them functions individually. Most of us are social animals, living in groups and interacting with others in a multitude of ways. Those range from pair formation and parent-child bonding up to being part of elaborate institutions like governments, corporations, stock markets, and organized religions. Having a greater understanding of mirror neurons may provide insights on why conflicts, ranging from friendly spats to global wars, arise. Surely there would be considerable value in understanding how such things come about and learning how to predict or prevent them.

The social sciences, as I've already mentioned, are not (at least so far) as rigorous as the so-called *hard sciences* like physics and chemistry. A physicist can predict what will happen under given conditions: a psychologist or sociologist, in general, cannot. Might they ever be able to? For example, might we someday be able to recognize conditions that *will* produce a war, and change them so it won't happen?

Isaac Asimov, in his Foundation series of stories,[5] imagined a future in which people could predict wars and other large-scale social trends, using a new science called *psychohistory*. He got the idea by analogy with a branch of physics called *statistical mechanics*. A gas, such as the air in which we all live our lives, consists of vast numbers (way beyond Carl Sagan's "billions and billions") of molecules dashing about in all directions and with a wide range of speeds. They travel in straight lines until they bump into something, but from time to time they do collide with another object (either another molecule or a container wall), and the collision causes each molecule to go off in a new direction at a new speed. Each molecule follows the basic laws of mechanics at all times, but trying to apply those laws to all the molecules to predict the behavior of the entire gas would be a hopeless job. We could never know the position and speed of every molecule at any time. Even if we could, there's no way we could do all the calculations required to figure out what they'll do next, even with any imaginable computer. But for many purposes, we don't need to. When people work with gases—for example, in designing a gas stove or a balloon—we don't care what each molecule is doing. We care only about macroscopic properties like temperature and pressure, which are determined by the aggregate behavior of *all* the molecules. Those we can calculate by treating the positions, speeds, and directions of the molecules as random, and applying the laws of statistics.

Similarly, while human populations consist of thousands, millions, or billions of individuals, each behaving in a far more complicated way than any gas molecule, we're often concerned with their aggregate rather than individual behavior. Business decisions and public policymaking need answers to questions like: What will the stock market do for the next year or ten years? Do we need to worry about a depression? How likely is a large-scale war in the next year, decade, or century—or another terrorist attack like 9/11? Does global warming pose an imminent threat to civilization, or even human life itself? Is there a way to get people, voluntarily or under coercion, to do the things that would be necessary to prevent or undo the damage?

Asimov's psychohistory is based on the idea that such predictions can be made by applying statistics to the behavior of large groups of individuals. For example, we may not be able to look at any man or woman on the street and guess how much of his or her income is likely to go into savings in a given year. But we may, if we have enough data on the correlation of savings with things like inflation rates and war threats, be able to make a reasonable estimate of how much the entire country is likely to save. Economists, of course, have been trying to do this for some time, but the results are still, at best, mixed.

Part of the problem is that the economic numbers may not tell the whole story, or even the most important part of it. Things can change radically and quickly if, for example, a charismatic demagogue starts preaching a philosophy of racial superiority that becomes wildly popular, as in pre–World War II Germany. Or if certain risky investment practices sweep the land, like margin buying before the Great Depression of 1929 or the dot-com boom and bust of our own recent past. But what causes some ideas or attitudes to sweep through large parts of a population, while others don't?

Some scientists in recent years have been struck by the parallels between genetics and the spread of ideas. Both involve the transmission and replication of information, with some units of information becoming widespread while others don't. Might it be possible to analyze the propagation of ideas in a way similar to that used for studying the propagation of genes?

In his 1976 book *The Selfish Gene*, Richard Dawkins suggested that it could, and introduced the term *meme* as a unit of cultural information analogous to the gene in biology.[6] The term has since caught on and become the germ of an embryonic science called *memetics*, by analogy with genetics. Those who are exploring the idea (see the References in the back of this book for some examples) think that the ways memes spread or fail to spread through societies may be describable with unprecedented precision—and predictive ability—by mathematical theories closely paralleling those of genetics.[7]

The analogy isn't perfect, of course. For one thing, the concept of meme is not, at least so far, as sharply defined as that of gene. A gene, as we have already seen, is now understood to be a specific sequence of DNA code that governs the development of specific biological traits and processes. A meme can be coded in a wider range of ways, such as slogans propagated verbally, songs, printed books, or belief systems propagated through several such vehicles. As the word is commonly used, a meme is actually more like a virus than a DNA sequence in a normal chromosome. Broadly speaking, it's an idea, but not just any idea. It's an idea that can be passed on from one person to others (in other words, it's infectious), taking on a life of its own and using those who hold it (its "hosts") as a means to its own future survival and propagation rather than its hosts'. As such, it can be either harmful to its hosts (like those memes that engender suicide cults) or beneficial (like the idea that observation and reason can be used to better understand and cope with the world).

Another critical difference between genes and memes is that genes propagate in a Darwinian way and memes in a way more like the scheme earlier proposed by Jean-Baptiste Lamarck for biological evolution. Lamarck, a nineteenth-century French army officer turned naturalist, deserves more credit than he now usually gets for his impressive contributions to the system of classification of plants and animals started by Carolus Linnaeus. Lamarck was, in fact, one of the earliest proponents of the idea of evolution. Unfortunately, he is now most remembered for the mistaken details of his evolutionary scheme. He thought that body parts that animals used often would develop disproportionately and those they didn't use would atrophy, and that those changes would be passed on to their offspring. Thus, for example, giraffes might have arisen by antelopes stretching their legs, necks, and tongues to reach tree leaves; with time, those organs would become permanently elongated. The next generation, born with that head start, would repeat that process, eventually leading to the extreme elongation we now see in the giraffe.

We now know that biological evolution doesn't work that way. As Mendel and many later scientists have shown, anatomical changes produced by an individual's environment and experience have no direct effect on its offspring. What is inherited are changes in DNA, which cause offspring to develop in a new way. Cultural information, however, is much more "Lamarckian": ideas—memes—that an individual learns by reading or going to church or protest rallies can be, and are, passed on to later generations. The result is that for "time-binding" animals like humans—that is, beings that can transmit what they have learned to others, including their offspring—there are two possible kinds of evolution: changing the characteristics of future individuals by passing on mutated genes, and changing the characteristics of the culture by passing on learned memes. And that memetic cultural inheritance can have the effect of enhancing individual abilities by teaching them skills or providing them with external aids like fire or computers.

If it is possible to develop a real science, or even a systematic engineering methodology, for memetics, it could prove valuable for such purposes as recognizing conditions that are likely to breed epidemics of terrorism and taking effective measures to prevent them. On the other hand, such knowledge might be used by a future Hitler type to spread his poison even more effectively. It could make advertisers even more effective at making people want products they don't need—and it could help individuals develop stronger "memetic immune systems" with which to resist such pressures.

Like any knowledge, this kind could be put to a wide range of uses, both beneficial and harmful, and if it becomes functional and widespread, an "arms race" between such uses is likely to develop. This scenario does not seem particularly imminent; memetics at present is, at best, in a very early stage of development. But we do need to keep the possibility in mind as something that we may have to deal with at some future (and not necessarily distant) time, particularly when you consider the ways it might converge with some of the other technologies already developing.

MINDS AND MACHINES REVISITED

I discussed earlier the sizable differences in how our minds and our computers work. I mentioned that each excels at particular kinds of work, and described an important and familiar example of how our ability to use computers was enhanced by making at least the most visible aspect of the computer's operation match more closely the way minds work. How far might that process go, and with what kinds of changes on both sides? So far I've assumed that computers will continue to function in essentially the same ways they have so far, with pure digital coding, precise memory storage in very specific locations, and the need to make the user interface more closely resemble the way users are accustomed to handling information. But what if the computers themselves could be made to function more like our minds?

It might seem that this sort of imitation would require detailed imitation of the circuit structure of the nervous system, but this may not be either necessary or sufficient. It's fortunate that we may not need such literal imitation, because it's almost certainly very difficult. I've already mentioned the complexity and ethical considerations of reverse-engineering brains; there's also the fact that no two brains are exactly alike. The brains of highly trained violinists, for example, contain many "extra" connections not found in the brains of nonviolinists. Making computers that can "think" more like us may depend less on the detailed imitation of the circuitry of any particular brain than on learning *how* brains do what they do and figuring out how to mimic those processes with hardware.

Can machines think at all? People have heated discussions about this, which often seem motivated more by a deeply ingrained need to believe they can't, than by a real desire to know the answer. Indeed, the answer may depend on what you mean by "think," and we can always redefine it to our own advantage—but a good case can be made that that smacks of cheating. If we can agree on what we mean by thinking, we ought to be able to figure out whether we can make a machine to do it—and maybe even go the next step and actually build such a machine.

Researchers in artificial intelligence (a term probably first coined by John McCarthy, meaning humanlike intelligence in a machine) often speak of the Turing Test proposed by Alan Turing.[8] If an interface (such as a keyboard) is set up allowing a known human being (you, for instance) to converse with an unseen entity, and you can't tell whether the other party is a person or a machine, then it is, for all practical purposes, intelligent. The interactive menus we all encounter far too often on telephone calls to businesses clearly don't pass the test. If you give a response that's not one of the ones they're programmed to recognize and respond to, they just say something like, "I'm sorry, I didn't understand that," and repeat the question.

More sophisticated programs along those lines give more convincing imitations but still don't really qualify. A famous program named Eliza does such a good imitation of a psychotherapist that it has fooled people; but it does so by "knowing" a large number of common questions and appropriate responses, and can be stumped by your going significantly outside that repertoire. It could be made harder to stump by giving it a larger inventory of questions and answers, including the ability to make conversational gambits on its own. You may object that it's still just doing what it was programmed to do—but I might counterobject that we are too, and the only real difference is how the programming was done.

But is that really true? Or is there something more that humans do that for some intrinsic reason machines can't? Many have suggested that the difference is that we can make connections between seemingly unrelated things to come up with something new—for example, a physicist idly watching the bubbles rising in his beer and being inspired to invent the bubble chamber, which became an important kind of detector for particle physics research. Or Newton watching an apple fall, wondering whether whatever made it do so was the same thing that held the Moon in its orbit, and creating his theory of universal gravitation.

Well, maybe. But as computers become capable of extremely fast computation, and storing huge amounts of information, is it not possible that they, too, could be made to search all that information for relationships and recognize ones of significance? I see no intrinsic reason why not; neither do scientists who have given the matter a great deal of thought, such as John McCarthy and Marvin Minsky. Some people find the possibility frightening, and indeed it warrants some serious caution. If you can build an artificial intelligence capable of concluding, "Cogito, ergo sum," might it not go on to conclude that it should serve its own interests rather than yours? Certainly a sobering thought, particularly if that intelligence also controls robotic appendages that can translate its thoughts into real-world physical actions.

On the other hand, might such a robotic intelligence instead become an interesting and likable fellow being, a colleague, even a friend? Here again is an area where science fiction can serve as a virtual laboratory for exploring the possibilities. It has already explored a wide range of them, from the early robot stories of Isaac Asimov[9] to the much more recent ones of Joan Slonczewski in which natural and artificial intelligences exist and work side by side, with refreshingly little concern for which are natural and which artificial.[10] It will continue to refine its visions as real-life researchers continue to refine theirs.

In the meantime, without passing judgment on the desirability of creating machine intelligence at or beyond a human level, I will briefly mention just one more line of research that may help bridge the gap: a kind of computer that does seem at least qualitatively more like our brains—sometimes to an almost eerie degree. It's called a *neural net* (or, sometimes a *neural network* or *neurocomputer*), and you don't so much program it as teach it.[11] Just as with human students, you don't know exactly what's going on in its "mind"—but it really doesn't matter as long as it gives you the results you want. And for some kinds of problems, it does that better than the more familiar kind of computer.

An important difference between our brains and most of our com-

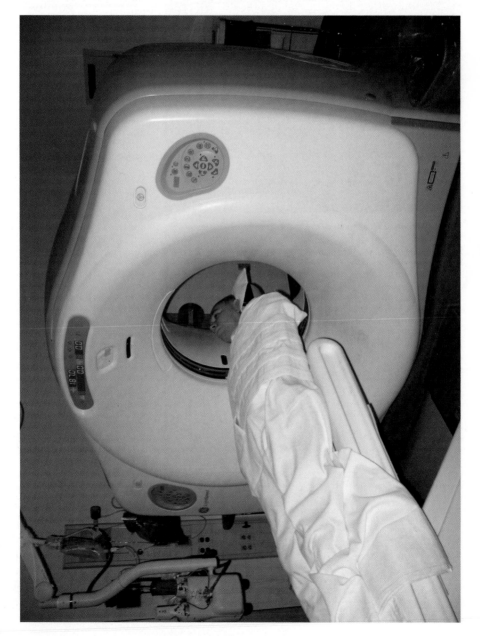

PLATE 1.
A CT scanner in use, combining x-rays and computer analysis to gain lifesaving information about a patient's internal health.
(Reproduced by permission of Henry G. Stratmann, MD.)

PLATE 2.
A convergence of aviation and big-building technologies sets the stage for a massive act of terrorism: the destruction of New York's World Trade Center and the deaths of thousands. *(AP Photo/Chao Soi Cheong. Used with permission.)*

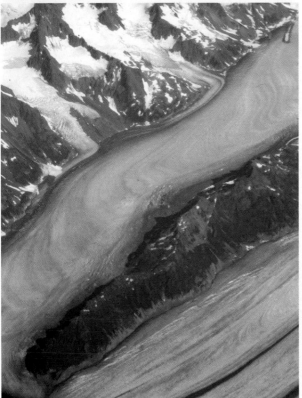

PLATE 3.
A natural convergence: when streams of water (rivers) or ice (like these glaciers in Alaska) converge, they form bigger, more powerful streams, carving ever-deeper channels into the landscape.
(Photo by the author.)

PLATE 4.
The evolution of electronic switching devices. *Left to right across top*: an early vacuum tube (from the 1940s), a later vacuum tube (from the 1950s), and a transistor (from the 1960s). Each of the three does essentially the same thing. At the lower right is a micro-chip, an example of large-scale integration from around 2000: it contains more than 37 million transistors (plus associated circuitry). (The chip itself is the small bright rectangle at the center; the larger square is just a holder for the chip and a means for connecting it to external circuitry.)
(Photo by the author.)

PLATE 5.
A pre-9/11 Manhattan scene shows several stages in the transformation of American urban landscapes by big-building technology.
(Photo by the author.)

PLATE 6.
This futuristic building, housing the Experience Music Project and the Science Fiction Museum in Seattle, was designed by Frank Gehry with essential help from computers. *(Photo by the author.)*

DNA Replication Prior to Cell Division

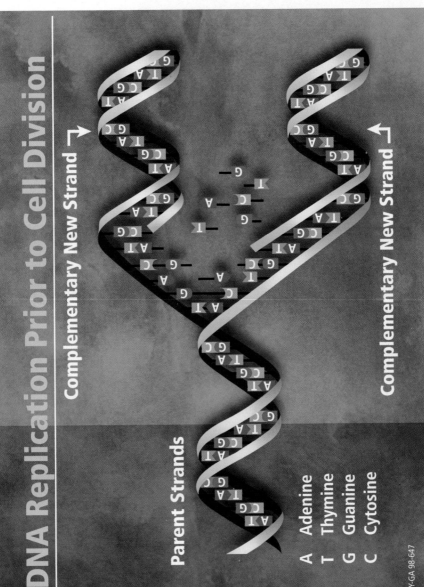

Complementary New Strand

Parent Strands

Complementary New Strand

A Adenine
T Thymine
G Guanine
C Cytosine

Y-GA 98-647

PLATE 7.
The central molecule of life. At the left we see the characteristic double helix of intact DNA; at right we see how it divides by "unzipping" down the middle and then each strand builds a new complete double helix by attaching complementary bases. *(Genome Management Information System, Oak Ridge National Laboratory [http://genomics.energy.gov]. Used with permission.)*

PLATE 8.
A planetary gear in a cutaway illustration, with atomic rows serving as gear teeth. Small gears (light blue) mesh with and roll between the input shaft (green) and the casing (gold), turning the output shaft (blue) at a reduced angular speed. This highly symmetrical design cannot be fabricated today, but can easily be modeled using standard molecular mechanics methods.
(Copyright K. E. Drexler. Used with permission.)

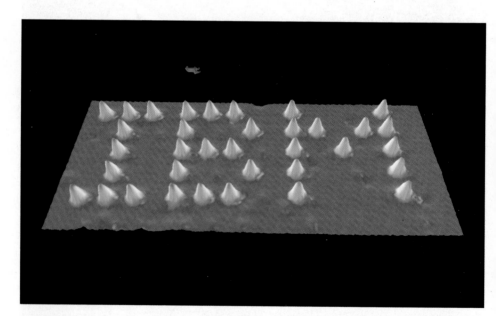

PLATE 9.
The initials "IBM" spelled out by thirty-five xenon atoms, each placed and imaged using a scanning tunneling microscope. *(Courtesy of IBM Research, Almaden Research Center. Unauthorized use not permitted.)*

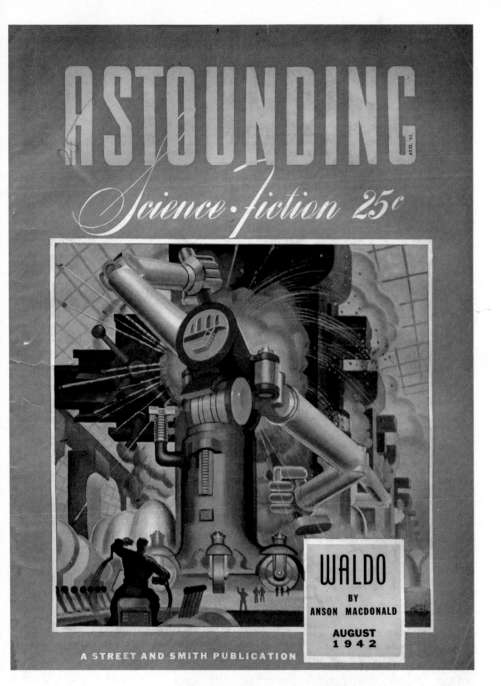

PLATE 10.

The first picture of a "waldo," an industrial manipulator that mimics the motions of its operator—now in common use, but receiving its name from this 1942 science fiction story in which it was first described. *(Astounding Science-Fiction. Copyright 1942 by Street and Smith Publications, Inc., reprinted by permission of Dell Magazines, a Division of Crosstown Publications.)*

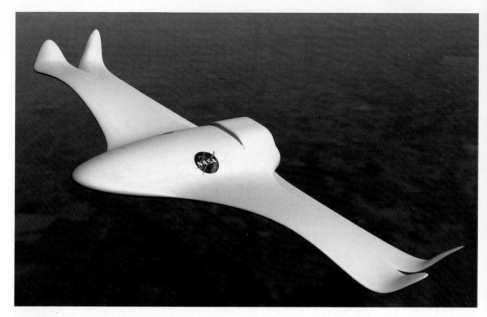

PLATE 11.
An artist's conception of a "morphing airplane," a NASA design incorporating several of the "smart technologies" mentioned in the text, including the ability to continually reshape its wings, much as a bird does. *(NASA Dryden Flight Research Center [NASA-DFRC]. Used with permission.)*

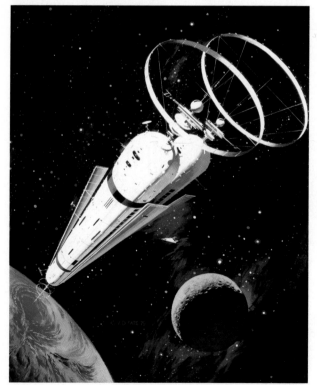

PLATE 12.
An O'Neill colony at L5, one of the especially stable points in orbit around Earth. This version incorporates two counter-rotating cylinders and holds what amounts to a small city. *(Photo by Vincent Di Fate. Copyright 2007. Used with permission.)*

puters to date is that we learn things by trying them repeatedly and getting positive reinforcement when we get them right and negative when we don't; while a conventional computer is simply told exactly what to do and then does it, very exactly. Another way to put that is that to solve, say, a math problem with an ordinary computer, you must already know how to solve it yourself. You tell the computer, step by step, exactly what it must do; its only advantage is that it can carry out those steps a lot faster than you can. But that approach doesn't work very well for every kind of problem.

Suppose, for example, that you want to recognize a particular face anywhere it appears in a stack of pictures, or the letter *A* wherever it appears in a book. You or I can do that easily, but it's extremely difficult to program a conventional computer to do it. You can't just tell it to compare what it "sees" with an image stored in a database and say it matches if a certain percentage of points in it are identical, and not if they aren't. A face might be seen head-on or in profile, from above or below its own level. The letter *A* may appear in twenty-four-point Olde English Script, nine-point Times Roman, or a six-year-old's scrawl; we recognize them all as being in some sense the same thing, but how do we do it? We don't know, in detail, and probably no two of us do it exactly the same way.

We have learned, though, that the human nervous system is a vast network of neurons connected by synapses, and that what kind of signal comes out of each neuron is determined by what signals it gets from other neurons. Furthermore, we use feedback. If we're trying to learn to recognize a letter in any of its many forms, we look at many things that are that letter and many that aren't. Somebody who has already learned to tell the difference tells us when we're right; and, in response, we apparently reprogram our neurons to respond differently than they did before. With repeated trials and readjustments, our recognition accuracy gets better and better. Eventually, when we've adjusted enough of them just so, we reach a point where we can almost always recognize anything that was intended as an *A*, and hardly ever misidentify something that wasn't. Where is the memory of what constitutes an *A* located? It's distributed through the entire network, and expressed in the way parts of it influence and respond to one another.

A neural net is a shamelessly analogous network of simple computers, acting like neurons, with interconnections corresponding to synapses. Just as the parts of our nervous system associated with vision get input signals from nerves connected to the eye, a neural net might get inputs from a camera. Each node is programmed to give an output determined by the signals it receives from earlier ones. For example, it might "fire" if and only if the sum of the signals it gets from two other nodes exceeds a certain value; or if it gets a signal from one but not the other; or if it receives signals from both at once. One of its signals—feedback—might come from nodes farther along in the system. The whole network can be set so that producing an output in a certain range is "right" and any outside that range is "wrong." After each trial, all the neurons readjust the weight they give each of their inputs, striving for a state where they give a "right" output a very large percentage of the time. In essence, it learns to recognize a type of pattern by being shown lots of examples and being given appropriate positive or negative reinforcement, depending on its responses.

Just like a human student of almost anything.

This type of computer has proved itself well suited to some types of problems, such as pattern recognition, and very poorly suited to others. Nobody would want to go through an elaborate and unpredictable "teaching and learning" process to get a computer to figure out how to do a straightforward number crunching problem, for which an exact algorithm can be easily written. But number crunching is a cumbersome way to do many of the problems of everyday life, in which things are not clearly this or that, but "sort of this" or "somewhere between this and that." That's the kind of process at which we excel, and the effort to teach machines to do it has spawned a whole new field called *fuzzy logic*— which, despite its seemingly oxymoronic name, has plenty of practical applications, from automatic cameras to robotic vacuum cleaners.[12]

There are clearly advantages in knowing how to make a (neural-net-based) computer using what we know about our own thought processes. The conventional digital kind has equally clear, but different, advantages. Quite likely our future will include both types, and some single pieces of equipment will incorporate both—and perhaps others that I can't describe because they use things we have not yet learned, about how we think, learn, and know.

CHAPTER 8

THE EXPLOSION IN INFORMATION TECHNOLOGY

We have seen up to this point that the handling and use of information, broadly defined as communication and computing, has played a central and pervasive role in shaping today's world. Its importance will continue to grow in the future. So far we have seen several streams converging to form "rivers" such as the telephone, photography, radio, movies, television (a kind of radio with paired signals carrying sound and pictures), and computers. Miniaturization—the art and technique of making things very small—has converged with all those streams to revolutionize how we do almost everything. And now all those rivers are flowing together to form something that has never before been seen outside science fiction. If so far we've been watching streams converge to form rivers, and rivers to form bigger rivers, we now seem to be entering a delta where they all empty into an ocean. There are deep waters out there, and they extend farther than we can see.

Until very recently, radio, television, telephones, computing, sound recording, and publication of the written word were recognizably distinct and separate. Even though they touched and interacted here and there, we still had radio stations that just did radio broadcasts, television stations that just did television, telephones that were just telephones, computers that computed, and publishers that just produced paper books and magazines. Now e-mail has become perhaps the most prevalent form of communication, and travelers listen to their hometown radio stations from the other side of the world via the internet. Many publications have electronic versions, and some have abandoned print altogether. Most cars warn their owners when they're developing internal problems, before those problems get bad enough to interfere with drivability. Some give them turn-by-turn directions on the road. Millions of individuals carry pocket-sized devices that combine not just two or three of the functions listed above, but several or even all of them—and more that I haven't mentioned yet. Some of the more advanced ones available as I write this (which many will consider obsolete by the time you read it) can be used interchangeably as computer, wireless telephone, radio, sound recorder, audio and video player and library, camera, photo album, e-mail and internet terminal, navigation aid, and emergency locator.

An essential part of what has made this possible is the amazing miniaturization of electronics, helped along by supporting technologies like the liquid crystals that make flat-screen displays possible. But there are other factors too. To begin understanding how we've gotten to where we are and where we're likely to be headed, we must pause to take a look at two more major currents that are playing vital roles in the mix, one very old, one almost new, but both commonly thought of as "futuristic": rockets and lasers.

ROCKETS, SATELLITES, AND WHERE AM I?

Rockets in their most basic form go all the way back to ancient China and India, where they were used in fireworks shows as part of religious festivals and, in a limited way, as weapons. They continued to play such roles off and on through ensuing centuries, but were too unruly to become much more.

A rocket is, in principle, the simplest imaginable kind of engine. Gas is heated in a closed chamber with one opening, and when the hot gas (the propellant) shoots out that opening at high speed, the chamber (along with whatever it's attached to) is accelerated in the opposite direction. The propellant may be heated in a variety of ways. Sometimes the propellant has been steam heated by a flame outside the chamber. But usually, in the things commonly thought of as rockets, the fuel is burned right in the chamber and the resulting exhaust gases themselves are the propellant. Since the gases are very hot and expelled very fast, the rocket can be accelerated very fast and its direction of travel is very sensitive to small influences such as uneven burning. Early rockets veered off course too easily to be accurate (or safe) as weapons. Stabilizing devices such as longitudinal rods trailing behind the rocket helped, but were not enough.

Around the turn of the twentieth century, the Russian physicist Konstantin Tsiolkovskii had bigger dreams. He wrote of the possibility of using rocket-propelled vehicles for space travel—for example, to go to the Moon, and even to build space stations and colonize the solar system.[1] Few took him seriously, but his was the first serious theoretical work on rocketry.

A bit later, the American physicist Robert Goddard put Tsiolkovskii's work into practice, building and launching an impressive series of experimental rockets (and earning, like Robert Langley, ridicule from the *New York Times* and neglect from the US government).[2] During World War II, Germany, forbidden by a World War I treaty to use conventional artillery and inspired by Goddard's work, developed the V2 rocket as a long-distance missile. It had better stabilization than most of its predecessors, but lacked remote guidance capability, so it did not have sufficient accuracy to play a major role in the war. However, it was important as an ancestor of later rockets, and in fact was the basis for the Redstone rocket later used in the early American space program.

During the 1930s and 1940s, even though most people dismissed the idea of space travel as "just science fiction" (as if that were equivalent to "impossible"), many science fiction writers were seriously exploring the possibility of lunar and interplanetary travel in their stories. It is in fact very difficult, in large part because the distances involved and the energy requirements are huge. But some scientists had made a good case that it was nonetheless possible, and fiction writers put faces on the results of their calculations. Those fictional adventures almost certainly played a significant role in making the real space program happen. Bright young people reading those magazines recognized the ideas in them as real possibilities, thought "I could help make that happen," and grew up to do so.

The process was not as smooth as it sounds, though. During the Cold War after World War II, the United States and the Soviet Union spent many years in the grip of fears of mutual destruction. Both developed weapons, including long-range guided missiles, which were essentially rocket-powered, remote-controlled bombs (another clear convergence of multiple technologies). In the early 1950s *Collier's* magazine published a series of illustrated articles about how existing or easily imagined technology could allow the exploration and colonization of space. Yet even as late as that, most people—even literate, well-educated adults—still scoffed at the idea, until October 4, 1957.

On that date, much of the country was shocked to hear that the Soviet Union had launched a basketball-sized satellite called Sputnik I into orbit around the Earth. It was partly a matter of national prestige and partly a matter of fear: if America's archenemy had a space capability that America didn't, wouldn't that put us in a new kind of danger? Russian satellites could spy on us, or act as orbiting weapons platforms, regularly passing over our country, ready to launch on command . . .

Those fears were not entirely groundless, and they were strong enough to serve as a goad to action. America started a crash program to develop the capability to launch its own satellites. Suddenly there was a great push to soup up American education in science, mathematics, and engineering, and to urge bright young people into those fields. The charismatic young president, John F. Kennedy, set forth a national goal of putting a man on the Moon and bringing him safely home by 1970, "not because it's easy, but because it's difficult."

Because of the difficulties already mentioned, overcoming them required yet another set of technological convergences. To launch something into orbit requires accelerating it to a speed of about 5 miles per second, or 18,000 miles per hour; to escape from Earth's gravity requires 7 miles per second, or 25,000 miles per hour. To achieve those speeds without destroying the objects being launched, chemists had to develop

fuels that could release huge amounts of energy very quickly; metallur-
gists and rocketry specialists had to develop materials and engine designs
that could contain such violent reactions; and electronic engineers had to
add control systems to make sure those expensive devices would do what
their builders wanted.

But all this was done, with a heady sense of excitement and adventure about it (which has regrettably been lost, at least for a while). Both
Americans and Russians launched a wide variety of satellites into orbit,
some of them carrying animals and then people, and on July 20, 1969,
John Kennedy's goal was reached: Neil Armstrong and Buzz Aldrin
walked on the Moon (while millions watched on live television), and
came home to tell about it.

The good part about the fear provoked by Sputnik I was that it got a
lot done. The bad part, in retrospect, is that much of it was done for the
wrong reasons. The public bought into the idea of the Moon as the prize
in a contest: in many minds the idea was not, "Let's go to the Moon
because of what we can learn or do there," but simply, "Let's beat the
Russians to the Moon!" Once that was done, as far as many people were
concerned, the race was over and they lost interest. Space travel and
exploration continued, but without the popular support and funding it had
once enjoyed. Over the ensuing decades probes were sent to and beyond
other planets, revolutionizing in a few years what we thought we knew
about the solar system. Eventually the knowledge gained that way may
lead to radical changes in our lives. Space may become an important
source of raw materials, a better place to conduct industrial processes that
are impractical or dangerous to do on Earth, and even an additional home
for humanity (which could be vital in the event of a planetwide cata-
strophe such as a large asteroid impact).

In the meantime, closer to our present home, artificial satellites have
found many practical applications in everyday life. One of the most significant is the widespread use of communications satellites (arguably first
proposed by science fiction writer Arthur C. Clarke in 1945[3] and independently conceived by at least one other person, John R. Pierce of Bell
Laboratories).[4]

Radio communication—including television and some kinds of long-distance telephony—has the fundamental limitation that waves from the transmitter have to be able to reach the receiver. And they don't travel well through large amounts of matter. Relatively long waves, like those used by AM broadcast stations, can do a moderate amount of "bending" around obstacles. Shorter ones, like those used by FM and television stations, are pretty much limited to line-of-sight transmission. Amateur radio operators and some broadcast stations with international ambitions use intermediate frequencies called *shortwave* to transmit across long distances by bouncing the signal off the ionosphere, an ionized layer high in the atmosphere. But atmospheric conditions, including the height and degree of ionization of the ionosphere, vary so much that long-distance communication by that method is much better at some times than others.

Clarke, Pierce, and others realized that a similar but much more reliable way to get long-distance radio communication would be to have satellites orbiting high above the Earth that could receive signals from one point on the surface and retransmit them to others. If such a satellite is orbiting at just the right height over the equator, in an orbit called *geosynchronous* or *geostationary*, it makes one circuit in exactly the same time as the point on the surface directly below it. Thus it appears to stay fixed relative to the Earth. Such an orbit is particularly desirable for a communications satellite because ground stations can keep their antennas aimed at it without complicated and expensive tracking equipment.

However, geosynchronous orbit (GEO) is about twenty-two thousand miles up and necessarily right above the equator. So it's expensive to put satellites there, and they're not very useful for communication between surface points far from the equator and in the same hemisphere. It's cheaper to launch things into *low Earth orbit* (LEO), only four thousand miles up (and within about ninety minutes); and those orbits don't need to be above the equator. By putting enough satellites into such orbits at various inclinations, it's possible to ensure that any point on the surface is in "sight" of at least one satellite at any given time.

By now a great many such satellites, of both types, have been deployed and used for a wide variety of purposes. The most obvious benefits are the easy worldwide telephone service and international broadcasting we now take for granted. Some satellites, instead of relaying phone calls or broadcasts from one surface point to others, instead measure physical conditions where they are and send them back to Earth for scientific study. Others photograph and otherwise monitor conditions *on* the surface from their privileged viewpoint high above it, and send their findings back for uses ranging from military surveillance, to prospecting for minerals, to studying deforestation, desertification, and climate change. Satellites have tremendously improved weather forecasting by letting meteorologists actually see large-scale weather systems and watch them move and evolve.

One particular kind of satellite has a special kind of importance. The *Global Positioning System* (GPS) uses a set of at least twenty-four satellites crisscrossing the Earth in LEO at various inclinations so that any point on the surface can always "see" at least three of them. Each carries a highly accurate clock precisely synchronized with the others and broadcasts a distinctive signal. A special receiver on the ground, also incorporating a clock synchronized with the system, can determine which satellites it's currently receiving signals from. A fast, compact computer built into the GPS receiver (commonly called a *GPS unit*), using stored information about the orbits of all the satellites, can quickly calculate where those satellites were when those signals were emitted—and, from that, where *it* is now.

In other words, a GPS unit provides a practically instantaneous way to tell exactly where it is. This ability has some obvious applications that nearly everyone would agree are beneficial, such as navigation on land, sea, or air, regardless of weather; drawing maps of unprecedented accuracy; helping hikers make sure they don't get lost; and helping rescuers find them if they do get lost.

A PURE AND POWERFUL LIGHT

In a very qualitative way, a laser can be thought of as similar to a clarinet or trumpet, in which sound waves bounce back and forth between the ends of an air column in a tube. For sound of a particular wavelength (and therefore frequency), the reflected waves will be in phase and reinforce each other, causing a strong standing wave to build up. Thus the column "likes" to vibrate at a particular frequency. A small part of the energy escapes as a traveling sound wave of that frequency, and that's what we hear as a note.

What happens in a laser is not quite the same, because it depends on the peculiar kind of physics called *quantum mechanics* that describes events at the atomic and nuclear scales. But it's qualitatively similar enough to make the comparison tempting. The word *laser* started out as an acronym for "light amplification by stimulation of emitted radiation." Albert Einstein described the basic idea in 1916.[5] Charles Hard Townes and Arthur Leonard Schawlow of Bell Labs filed a patent application for a practical version using visible light in the late 1950s,[6] and Theodore H. Maiman built the first working laser at Hughes Research Laboratories in 1960.[7]

In a laser, light is reflected back and forth in an "optical cavity" filled with an amplifying medium between two mirrors. The amplifying medium can take several forms; for example, it can be a solid (like ruby) or a gas (like a helium-neon mixture). Energy is fed into the medium from outside, sometimes as electricity and sometimes as bright light from another source. This energy "excites" atoms in the medium. Quantum mechanics requires each electron in an atom to exist in one of several sharply defined energy states, and excitation raises one of those electrons to a higher state. The excited state is unstable, so the electron soon drops back to its original state. In that process, it emits a photon, a particle of light of sharply defined frequency and wavelength. Some of those photons travel in directions practically parallel to the axis of the laser, so they can travel back and forth between the mirrors. Sometimes one of them hits and excites another atom, so in a short time quite a number of them can be bouncing back and forth.

And a few of those can escape through one of the mirrors, which is designed to let some of the light hitting it pass through. Because of the way it's produced, this light has several unique properties. It's monochromatic (i.e., it's all very precisely the same color); it's coherent (all the waves are in phase); and, since it's all going in very nearly the same direction, the beam is very narrow and (in relation to the power input) very intense. All this is in sharp contrast to "ordinary" light sources like flashlights, which radiate light over a wide range of frequencies and a wide range of directions, with the result that the energy is thinly spread and not extremely bright.

Those properties have led to a wide range of uses for lasers, and changed them from expensive laboratory curiosities to routine parts of everyday life. With no particular effort I can think offhand of at least nine in my house and cars, and I've probably missed a few. Some of the earliest and still commonest applications use the sharpness and intensity of the beam. A low-powered laser makes an excellent pointer; a higher-powered one can be a surgical or industrial cutting tool; a still stronger one can be a military weapon (reminiscent of the ray guns of old science fiction) or a power source for spacecraft.

Since the beam is so narrow and sharply defined, it can be used for reading compactly stored information. One of the earliest such applications was the bar-code reader, which interprets those patterns of black lines of various widths now found on practically every package of just about anything. A considerably more sophisticated descendant of that is the whole family of optical disks used for storing sound, pictures, and all sorts of computer data: CDs, DVDs, Blu-Ray disks, and their kin. They all store digital information as patterns of bumps and pits on a disk to be scanned by a laser. DVDs store much more information than CDs by using smaller bumps and pits; Blu-Ray gains an additional advantage by using a shorter wavelength of laser light. Optical disks quickly and almost completely supplanted vinyl LPs and cassette tapes, which had been the dominant forms of sound recording for several decades, because of several advantages. Optical disks store more information in less space, are less sensitive to physical abuse, and have fewer moving parts in the playback equipment (which is therefore less prone to breakdown).

One information-related application of lasers (mentioned briefly earlier) is a particularly interesting example of converging technologies from a few decades past. In 1947 Dennis Gabor described a process called *holography* or *wavefront reconstruction photography*. In ordinary photography, light from an object is focused to form an image on a film or plate, which looks to an observer like the original object viewed from a single point (that is, with one eye closed, since it's two-dimensional).

In holography, instead of recording an image of the object, the same film or plate can record the interference pattern formed by light from the object mixing with light (called the *reference beam*) directly from the same source that illuminated the object.

That pattern, called a *hologram*, bears no resemblance to the object; it's just an apparently random pattern of swirls. But when it's illuminated by the reference beam alone, it acts like a special diffraction grating that "reconstructs" the waves originally reflected by the object. An observer intercepting those waves appears to see the object itself, fully three-dimensional. He can even move his head and look behind objects in the picture; there's no purely optical way to tell whether he's seeing the real object or the holographically reconstructed image.

It's an astonishing effect for anybody who's never seen one before. But essentially nothing was done with it for more than a decade after Gabor's work, because it was very difficult to make holograms with the incoherent light from ordinary sources. When the laser was invented in 1960, it didn't take long for other experimenters to realize that such a source of coherent light could make holography far easier, so it quickly became a hotbed of research. (Ironically, once the laser made it relatively easy to experiment with holography using coherent monochromatic light, it wasn't many more years before people found ways to make them with ordinary white light. Now we see them fairly often on splashy magazine covers and as hard-to-counterfeit logos on credit cards.)

THE PHOTOGRAPHER'S NEW "CANVAS"

Holography, a new kind of photography, leads us naturally to another new kind of photography that has been embraced so eagerly by so many people that it has already become the predominant form of photography. It's questionable how long film will continue to be used at all. Many see this as an unmixed blessing, but the situation may not be quite that simple.

Around 1970, a photographer friend gave me a scare by telling me that the silver essential to film photography was expected to run out by the early twenty-first century. Since photography played a sizable role in my life, I found this alarming: if silver became unavailable for film, what, if anything, would replace it? Or would photography itself die?

The answer began to assert itself in the 1990s, when the first digital cameras were made. Those devices looked superficially like film cameras, with lenses focusing an image of the object being photographed onto a photosensitive surface. The difference was that, instead of film coated with a photosensitive chemical emulsion, the imaging surface was an array of tiny electronic photosensors (commonly either CCDs [charge-coupled devices] or CMOS [complementary metal oxide semiconductor] devices). Each of those recorded the color and intensity of the light hitting it as a digital signal called a *pixel* (short for "picture element"). The total picture was made up of a large number of those tiny squares. If there are enough of them, the eye doesn't notice that granularity, just as it doesn't notice the grain in a film photograph unless it's enlarged too much.

The basic nature of digital cameras has stayed the same, but (as with most electronic technologies) their quality has increased and their prices decreased rapidly. The first digitals were very expensive (on the order of ten thousand dollars!) and their image quality far from competitive with much cheaper film cameras. Few people used them; to professionals they were interesting toys, maybe with potential for the future, but certainly not serious tools for their present work. As I write this, just a few years later, that has changed so dramatically that more photography is now done digitally than with film, both by professionals and by amateurs. Part of the reason is that equipment prices have become more competitive; another part is that the quality has improved so much, and the variety of options has been greatly expanded, with cameras from the simplest to the most complex and versatile now available. Not long ago a three-megapixel camera was about the best you could hope for, and its pictures could compete with film only up to about 8 × 10 inch size. Now six or seven megapixels are common; ten- and even sixteen-megapixel cameras are available, and even (depending on your budget and needs) relatively affordable. Those still can't give quite the quality of a good film trans-parency, but that's not a fundamental limitation of digital. Moore's Law applies here, too, and it's only a matter of time before digitals can meet whatever quality standard you require.

If picture quality is only as good as that on film, that alone is not a reason to change, with all the expense and hassle of buying and learning to use a whole new system. The reason for digital's swift ascent to dominance is a whole array of clear advantages it offers. No film or developing is required, and a single memory card can store far more pictures than any roll of film. The photographer can see at once how a picture came out, on a small LCD (liquid crystal display) screen built right into the camera (which doubles as a viewfinder for picture-taking and is preferred by many to the optical type). If she isn't satisfied with it, she can retake it until she is—and erase the ones she didn't like, freeing up the memory they occupied for reuse (unlike film, which, once exposed, is gone). Later on, the pictures can be manipulated—enlarged, cropped, brightened, made more contrasty, combined, and otherwise edited and manipulated—in a computer, far more easily than with film images in a darkroom.

But perhaps the biggest advantage of all, in many people's minds, lies squarely in the easy convergence of digital photography with all the other digital technologies surrounding us. Giving a copy of a film picture to someone else requires either making one or waiting to have someone else make one, usually for a fee; and while that's being done, you don't have the original. Giving a copy of a digital picture to someone else is as simple as e-mailing it as an attachment, with no need even to touch a physical picture. Or you can post it on any of several Web sites where it's available for viewing by anybody interested (and downloading, by anybody interested enough to pay a fee).[8] Since the picture data are stored digitally, they can be copied with no loss of quality. They don't even have to be recorded by a "camera" in the first place—which leads us to yet another of the main ingredients in the digital stew now boiling around us.

THE DIGITAL MELTING POT

In the last very few years, many of the traditional boundaries between different information technologies have been breaking down, with single devices offering almost any imaginable combination of them in a single package. Shortly we'll look at some of the many opportunities—and problems—offered by those multifaceted combinations. But first we should consider a couple more key players that, at least in their first incarnations, combined just two or three "separate" technologies.

One of those, and a profoundly influential one, is the cellular telephone.[9] Decades ago comic strip readers watched Dick Tracy get out of jams with the help of his wrist radio. Few ever expected to own one, but now many—if not most—people routinely use something very much like it, in function if not method.

It isn't practical to wear a long-distance two-way radio on your wrist, for reasons we already looked at in connection with communications satellites. But it is practical to build a network that works somewhat like a satellite system, on a more down-to-Earth scale. A cell phone is a radio transceiver, but it doesn't have a very long range. It doesn't need one—all it has to do is send signals to, and receive them from, a nearby relay tower. That tower can then use some combination of stronger radio transmitters and telephone land lines to get the original signal to another relay tower close to the intended receiver—whether another cell phone or a traditional phone. Naturally it takes computers to keep track of all those signals and route them so that they get where they're going efficiently, clearly, and without mutual interference. And cell phones commonly incorporate small computers used for such tasks as storing contact information—in effect, a small phone directory—for people the user might wish to call or be called by.

The other device that is revolutionizing how we do many things, and combines easily with digital photography, is the digital audio player. This is now most often referred to colloquially as an MP3 player (even though MP3 originally meant one of several specific data storage formats) or an iPod (even though that is a registered trademark for a type made by Apple Computer). CDs and DVDs caught on quickly because they brought the advantages of digital storage to fields formerly dominated by phonograph disks and magnetic tapes. But computers were already using other kinds of digital storage media with still greater advantages, and any digital medium can store any kind of digital data. So it was hardly surprising when two very compact, high-capacity media associated with computers began being used for music: magnetic disks and flash memory.

Magnetic disks first appeared as floppy disks, spinning disks (usually packaged so their disklike nature was not obvious) on which information was encoded as patterns of tiny regions that were magnetized or not magnetized. Those were read by passing over a "head." Floppies held only small amounts of data, by present standards (800 kilobytes, for the first ones I used), so it was often necessary to eject one and insert another, and they simply couldn't handle files of the sizes required for high-quality graphics or sound recordings (typically several megabytes). Those were major reasons why floppies were soon largely superseded by hard disks (or hard drives), which worked on the same principle but stored data much more compactly. Thus much more of it could fit on a single drive. (The one on the computer I'm using to write this, quite modest by present standards, holds sixty gigabytes, equivalent to seventy-five thousand of those floppies I started out with.) Over the ensuing years hard disks have continued to become much smaller and much more reliable, and now quite tiny ones can be built with plenty of room for music and pictures.

Flash memory (so far) can't hold as much data in the same space as a hard drive, but it can hold a great deal—and has the significant virtue of having no mechanical moving parts. It's simply a solid-state circuit whose configuration can be set by simply feeding digital signals into it, and then retained without being connected to a power source. Many computer users use "flash drives" (which aren't really drives at all, since nothing is being driven) as an almost ridiculously easy way to back up and transport important files. With sizes now on the market typically holding a few megabytes, they're not big enough to back up all the application software a computer might use; but they can easily store substantial numbers of documents, pictures, or songs. They're also very forgiving and durable. People often carry them in pockets or on keychains, and a friend of mine accidentally ran one through a load of laundry and found that it still worked fine.

Some audio players use small hard drives; others use flash drives. Each type has its own advantages and disadvantages. Hard drives can hold more but are bulkier and more prone to mechanical failure. Flash drives can't hold as much but can be very tiny and have no moving parts to break or malfunction. Both types also include hardware and software for playing and otherwise manipulating the music stored on them. Their great advantage is that they can hold so much. A recorded music collection that would occupy many feet of shelf space on LPs or even CDs can all be packed into a single package that easily fits in a palm. Furthermore, any of it can be accessed quickly and easily, or arranged at the user's whim into customized programs playing any desired combination of pieces in any desired order.

And where do they get that electronic library? Most often by downloading it from Web sites devoted to that purpose—sometimes legally, sometimes not.

Since so much memory and processing power could be packaged so compactly, it quickly became evident that cell phones and digital players did not have to be limited to their original purposes. Digital audio players also became video players and gaming machines. Many cell phones soon came to incorporate digital cameras, which could be used not only to take pictures practically anywhere, but also to instantly send them practically anywhere else. GPS units also became standard components, enabling emergency services to home in on the caller's exact location without having to try to get directions from someone who may not be able to give them.

And, of course, people have always been ingenious about finding new uses for devices besides the ones for which they were intended. As I write this, many schools have been banning cell phones and iPods in classrooms where tests are being given, because too many students have learned to use them as high-tech crib sheets.

TRANSFORMATIONS AND CHOICES

With all those information technologies essentially merging into a unified whole, with the boundaries erased or smoothed over, both individuals and organizations are rapidly gaining unprecedented power. Any powerful tool can be used in many ways, some of which almost everybody would consider good, some most would agree are bad, and many that are debatable. No technology is intrinsically and purely good, or intrinsically and unmitigatedly evil. Once we have abilities, we have to make choices—both private and public—about what to do with them. Since some people are much more interested in personal gain than fairness, some regulation is likely to be necessary as "wars" between competing interests break out. And that doesn't just mean little tweaks to existing standards of law and manners, but sometimes major revisions of how we think about those matters.

I hinted earlier at a couple of examples: illegal file sharing and use of cell phones and digital players to cheat in school. But those are just the tip of the iceberg. Consider, for example, just some of the many ways computers and the internet have transformed societies, creating both new opportunities and new problems that must be dealt with.

English teachers sometimes lament the very informal nature of e-mail and text messaging, but they must also admit that those practices have revived interest in written communication with a vigor not seen in a long, long time. The great thing about e-mail is that it's so easy; and the terrible thing about e-mail is that it's so easy. It tempts people to "flame" when diplomacy would be more prudent, and to write (and expect others to read) vast quantities of stuff that they wouldn't bother with if it were just a little more trouble. That effect may fade at least a little as the novelty wears off, but we're still in an age of constant innovation, so it may take a while—if it happens at all.

We do not all see the same things as problems. I occasionally hear people complaining that others who use e-mail and frequent chat rooms are "interacting with machines instead of people." This one, after just a little thought, is hard to take seriously. It's very much like complaining that somebody who spends a lot of time on the phone is interacting with a handset rather than a human. The internet, like the phone, is not a *substitute* for human interaction; it's just another way of connecting.

On the other hand, the very ease of electronic communication has spawned a whole complex of other problems that most of us can agree on. It's almost as easy to e-mail something to vast numbers of people as to one, so most of us who use e-mail at all spend too much of our time dealing with spam, unsolicited junk from people we don't know and would rather not know. Computer manufacturers concoct "spam filters" to try to stem the tide, but they're not infallible. Sometimes they mistakenly block messages we want or need to get. Spammers invent ways to circumvent whatever filters we have in place—which forces the "good guy" programmers to develop better filters.

And so on. It is, essentially, a classic example of an arms race.

And it's not the only one. The fastest-growing crime in the United States is one for which no name existed (at least in common use) just a few years ago: identity theft. Psychologists talk about, and treat, a new disorder called *internet addiction*. And some malicious programmers' designs go way beyond abusing somebody else's credit. Some write viruses, programs that can spread via the internet to many private and public computers. Their primary purpose is destroying the information stored in those computers and making them unusable—sometimes just for the fun of it, sometimes as a terrorist weapon.

Another significant transformation is the way computers, both singly and in concert, have radically altered the way publishing is done, and forced writers, performing artists, publishers, legislators, judges, and lawyers to rethink what *copyright* means. That process is far from over. It started with desktop publishing: the use of small in-house computers to do much of the work of getting from an author's manuscript to a finished book or magazine. When I started editing *Analog*, authors submitted typed manuscripts, a copy editor wrote corrections and detailed typesetting instructions on them, and we sent them off to a different company to be typeset. Even then, that no longer involved physical type; but it did depend on someone's retyping the entire manuscript—and quite possibly introducing new errors in the process.

Now even the meaning of *manuscript* has changed. Most editors I know still prefer to read paper manuscripts in deciding what to buy; but once we purchase something, we assume the author will provide an electronic version that we can use for production. Having that streamlines the whole process and makes it more accurate. Nobody has to retype the whole thing, so it all gets done faster, and there's less chance for errors to be introduced. In effect, copy editing and typesetting have been combined into one operation, with the most treacherous step eliminated.

But the end product is still a paper book or magazine, though it may also be offered in electronic formats.

However, some publishers have gone even further, in various directions. Some publishers still produce bound paper books, but instead of printing and storing thousands of copies and hoping enough sell to make it worthwhile, they store only an electronic copy of the finished book and print and bind a copy only when a customer wants to buy one. One such "print-on-demand" publisher told me a few years ago that it took him about twenty minutes to make a book to order. I suspect it's faster now, and this system has become a popular way for authors to keep their out-of-print books available for sale.

Some publishers don't produce paper at all, but publish only online. Finding ways to make money at this has required creating new business models, but some have managed and the practice will likely increase. In fact, it's so easy to publish online that essentially anybody can do it, and there has been a great increase in authors publishing their own work. The blog, or Weblog, is arguably the first new literary form of the twenty-first century, and some blogs have attracted considerable followings.

Another candidate for "new literary form" is the massive and continuously evolving collaboration. One of the first places this appeared was in computer programming, which is usually done within corporations and the details are closely guarded secrets. But one of the stablest operating systems, Linux, evolved as a cooperative effort with all the code posted online and subject to change by anybody.[10]

Another example of massive collaboration and open-source development, perhaps of more general interest, is Wikipedia, a free encyclopedia that exists only online.[11] Hypertext (with illustrations) is, in meaningful ways, the most logical medium for an encyclopedia, because so much of what one learns from such a reference is found by starting with one article and following cross-references. I can say without exaggeration that I got a major part of my education that way, but for most of my life following cross-references meant pulling out another volume and leafing through it to the target pages. With hypermedia, the process becomes much simpler and faster. You can get right into a reference by simply clicking on a link—and the illustrations can include not only higher-quality pictures than a book could include, but also animations and sound.

The entire internet—with browsers like Mozilla Firefox and Safari to navigate through it and search engines like Google to find relevant information—can be thought of as the world's biggest and worst-edited encyclopedia. There's an enormous quantity of information on it, some of it excellent and some of it utter bunk, and there's no obvious way to tell the difference. Wikipedia, started in 2001, is a deliberate attempt to establish a section of the internet that can actually be treated as a useful and reliable encyclopedia, continuously evolving through the input and mutual corrections of thousands of contributors.

Many (including me) were skeptical of the idea, and indeed it must be approached with some caution. It doesn't have official editors checking contributions, contributors are not required to be experts, and cranks or pranksters can sabotage articles by deliberately introducing errors. However, enough real experts have chosen to participate that errors tend to be corrected quickly. Several studies have attempted to quantitatively compare Wikipedia with traditional encyclopedias, and found their accuracy and reliability surprisingly similar.[12] And Wikipedia does have the advantages of fluidity and practically unlimited scope, so that it includes information on a great many topics that have not yet found their way into conventional encyclopedias.

Not surprisingly, the extreme ease and cheapness of online publication has in some cases led to misappropriation and unauthorized publication—and heated debate over what rights people have to publish other people's work or to forbid such publication of their own. Copyright was created to ensure that creative artists would have a chance to earn reasonable compensation for their work by forbidding others to profit from it without the creator's permission. But copyright law was written before the internet, when a private individual was unlikely to make enough copies of an article or record to seriously threaten an artist's livelihood. That changed dramatically when it became easy for practically anyone to send essentially unlimited numbers of copies of just about anything to others, who would then have little incentive to buy copies that they otherwise might have bought.

Cases like the early days of Napster, a Web site set up to facilitate such file sharing of songs, led to heated controversies over just how, if at all, copyright law applied to such situations—and how, if at all, it needed to be changed. As it stood, it didn't seem to apply. Since the files weren't being sold, there was no clear violation of the law as written and heretofore enforced. Yet for the first time in history, privately made, freely distributed copies could be distributed in such huge numbers that they were serious competition for legitimate sales. Avid users of Napster defended their actions with catchy but semantically empty slogans like, "Information wants to be free!" while ignoring the question of what would motivate professional-quality artists to continue to create if they couldn't even hope to make a living from it. Copyright laws were not created out of a purely altruistic concern for artists, but to encourage them to do work that benefits the rest of us, and to enable corporations to profitably support that work by publishing it.

Napster lost a court battle, but later returned, along with several similar sites, with a mechanism for charging its users and passing a share of the proceeds on to the creators of material being downloaded. But illegal exchanges still go on in other places, and so do the debates of how artists' rights, consumers' desires, and technological realities should (and can) be balanced in the future.

ELECTRONIC BREAD CRUMBS AND BIG BROTHER

Publishing is not the only area in which new technologies have wakened new controversies over rights and responsibilities. Remember the fairy tale about Hansel and Gretel, the children who laid down a trail of bread crumbs so they could find their way back out of the woods, but got lost anyway because the birds ate their markers? What if they had instead been able to put down things not only unappealing to wildlife, but also capable of calling out to them as they passed, "Hey, I belong to Hansel and Gretel"?

Well, we now have at least a couple of things along those lines. The kids could have made good use of them—but so could the witch.

One of them we've already mentioned. GPS has already proved itself an extremely useful tool. If you rent a car in an unfamiliar city, you can now get one that includes a GPS system coupled with a computer and database that will tell you exactly how to get where you want to go, without ever needing to unfold and refold a map. If you want a paper map, you can get a more accurate one than ever before because GPS was used in making it (and a computer may well have done the actual drawing). If you have a car accident and you don't know where you are, just dial 911; your phone will lead the police right to you.

On the other hand, it can lead the police right to you for any other reason, too—not a problem if you're in a perfect free society, but potentially a very serious one if you're in the jurisdiction of a corrupt regime and have political enemies. Rental car companies have been known to secretly plant GPS transponders on their vehicles and then charge the renters for speeding even when they were not stopped by the police.[13] Granted, renters agree to comply with traffic regulations, but should rental companies be playing policeman and secretly spying on their customers? Some parents routinely use GPS to track their children's every move. Does this interfere with the children's developing any sense of trust and responsibility? Does it condition them to accept such monitoring in adulthood, from governments or employers? Such questions may seem unnecessarily alarmist, but we're still in the early stages of those developments, and they warrant serious consideration before we go any further.

Another kind of electronic breadcrumb is one I haven't mentioned yet: the *radio-frequency identification* tag, or RFID (which some pronounce "arfid").[14] This was originally conceived as a more sophisticated version of the bar-code system, but it can do much more. A bar code has to be individually read by passing it directly in front of a reader. An RFID does not.

An RFID is a small object including a microchip on which digital information is stored, and a radio transponder and antenna. A reader, which may be some distance away, "pings" it with a microwave signal, and it responds by transmitting back to the reader the information it carries. That includes at least the identification of the object to which it is attached, and potentially more. The reader can then compare that information with data stored in a computer to determine, for example, the current price. It may also use that information to do things such as giving you coupons based on what you've bought in the past, and updating the store's inventory so the manager knows when to reorder. Unlike a barcode reader, an RFID reader can read everything in a grocery cart or an armload of library books at once.

RFIDs are not limited to acting as cash registers and keeping track of inventory. You may be more familiar with them as tags allowing you to breeze through expressway and bridge toll booths without stopping, or to buy fuel without handling paperwork or credit cards, or to get into the high-security building where you're employed. Some countries already have them in their passports and/or currency, and more plan to do so. Having them in passports can expedite your passage through immigration—or enable a terrorist with a reader to identify your nationality and pick you as a target. Having them in money makes it much harder to counterfeit—and makes cash transactions as easy to trace as those done by check or credit card.

In other words, with RFIDs in widespread use—and they're already well on their way—privacy will be very hard to come by. That may be a worthwhile trade-off, but not one that should be made without careful consideration. How much freedom are we willing to trade for how much security? It's an important question, touching the very foundations of our society, and we'll need to try to answer it with thought for the long term as well as short term. Many of the same people's answers would have been very different in August 2001 than in October 2001, but we'll be stuck with whatever laws we get for much longer than a couple of months.

CONTINUITY

All new technologies present mixed bags of opportunities and problems, but there's one problem that's peculiar to information technology—and may need more consideration than it usually gets.

With few exceptions, a car of new and greatly improved design will be rightly seen as an improvement by almost all buyers, even if it doesn't use any of the same parts as older models. A new typewriter doesn't have to be similar to old ones to be an improvement; each one produces words on paper, to be read by eyes, and the old and new are independent. The same is true of cameras, as long as all you want is to make prints: you can look at prints from a digital camera the same way you look at those from an old or new film camera. Radio and television are no problem: better reception and newer features are clear improvements, as long as you're watching them in real time (except for the inconvenience and expense of buying new equipment if stations switch to methods incompatible with old receivers).

But when you're concerned with storing information for later access and use, things get much more complicated. It's easy to get so caught up in the excitement of new and clearly better media that you forget the basic purpose of information storage. If, for example, you're trying to maintain a complete and ever-growing history of an institution such as a university, it does little good to adopt a snazzy new medium for archiving the latest records if by so doing you become unable to read early ones— thereby, in effect, throwing them away.

Yet in recent decades we've been doing exactly that to an alarming extent. I've seen several articles about the problem the National Archives has with a rapidly accumulating collection of materials stored in old computer formats that could be read only by machines no longer built or maintained.[15] Many of us have experienced this problem on a more personal level with such things as collections of music on LPs or audio-cassettes for which our old players no longer work and it's difficult or impossible to find new ones. If the collection is large, it may be prohibitively expensive in money and/or time to replace it all with the same content in newer media such as CDs or MP3s—even if it were available that way, and much of it may not be. This won't matter to you if all you're interested in hearing is the latest passing fad; but it will if, for example, there were songs that had special meaning for you in your youth and still do decades later.

There are hopeful signs that some people have begun to think about this problem and even to do something about it. Some of the more recent media actually have a degree of backward compatibility built into them. At least some DVD players, for example, can also play the earlier CDs.

It has also become easier for different kinds of computers, such as Macintosh and Windows systems, to talk to each other. But what we really need, if we want to preserve access to older but potentially valuable information, is to develop the mind-set of thinking about how we're going to do that even as we design new equipment. The more thought we give up front to how we're going to keep information stored in old media accessible as we convert to new ones, the less we're likely to lose and the more continuity we can have between past and future.

Of course, obsolescence (planned or accidental) is not the only way we can lose big chunks of our past. Another threat is physical disaster, either natural or human-caused (such as war or terrorism). Here the digital technologies we already have or can anticipate have special vulnerabilities. A collection of photographs and books can be destroyed in a fire, but there will usually at least be a little time in which human effort can save some of them. An equivalent collection of digital photographs and text files, on the other hand, can be completely destroyed in a fraction of a second by a single electromagnetic pulse from a bomb or a machine built by terrorists or pranksters. In fact, such a pulse can destroy *all* such collections in quite a large area.

We can take some precautions against such losses, by such means as keeping multiple copies of records in widely separated locations, at least some of them in heavily shielded vaults. But even if the records themselves are not destroyed, they can be made unusable if the technological infrastructure on which they depend collapses. It's all too easy to imagine that happening in a war fought with weapons available even today. If civilization collapses worldwide—not a clear and present danger, but certainly possible—well-protected paper books and film photographs may still be available to help rebuild it. Equivalent material that was stored only as digital files will be useless.

None of this is to suggest that we should avoid adopting such technologies; their advantages are too big to pass up. Hard copies of books and pictures have the advantage of not needing an elaborate technological infrastructure to be usable, but no hard copy can really be equivalent to a digital database or photograph in terms of the ease of finding, correlating, and using the stored information. Those capabilities are essential to the unprecedented strides now being made; but even as we take advantage of them, we need to concurrently think about how we're going to ensure continuity.

The Polish-American philosopher and scientist Alfred Korzybski characterized our species as the "time-binding" animal: the one that can transmit knowledge from generation to generation at an ever-accelerating rate, because we can learn from all past generations, not just from our own experience and that of our immediate predecessors. To appreciate how true that is, you need only think back over the stories in the earlier chapters of this book. Many accomplishments of our civilization could happen only because of individuals building on the work of many generations of predecessors working in different fields.

But that can happen only if we have access to what was done before. There are documents written hundreds or thousands of years ago that can still be read by people living today. There are others written just a few years ago that are already essentially unreadable. We'll need to take special care to make sure that in building history's most advanced civilization, we are not also building its most ephemeral.

HOW FAR CAN IT GO?

We have seen here some of the ways new information technologies have been transforming the ways we do almost everything in the last few years. I say "have been transforming" rather than "have transformed" because the process is far from over. Indeed, the changes in the next couple of decades will likely dwarf those of the last couple. We can easily imagine, for example, a future in which human-machine interfaces have gone far beyond "natural language" to "internal language." People might wear surgical implants that interface directly with the nervous system and provide full-time immediate access to a worldwide network of information sources and processing resources. It would be similar to the present internet, but "much more so."

Such a situation would, like all its predecessors, offer both unprecedented opportunities and unprecedented dangers. Clearly, such developments will depend on still more convergences of the lines of development we've already discussed. Those implants, for example, will depend both on still more miniaturization and on still better understanding of how both nervous systems and computers work. Continued miniaturization itself may depend on such convergences. For example, one mechanism that has been suggested for building still smaller computers is using DNA as a computing medium. Still others involve *quantum computing*, in which the spin states of individual atoms or electrons act as the basic "circuit" elements; or *plasmonics*, using light to produce electron density waves along the interface between a metal and a dielectric.[16] Just how much can be done will depend on how far miniaturization can go—how much capability can be squeezed into how small a package.

We'll take a closer look at some of those mind-stretching but very real possibilities in later chapters. First, though, we must consider one more current that promises to be one of the dominant forces in the coming convergences—and in fact, has already begun to fulfill that promise.

CHAPTER 9
NANOTECHNOLOGY

The fundamental difference between nanotechnology and all previous technologies is simple: nanotechnology (in its strictest sense) makes things from the bottom up; everything else is top-down. That is, nanotechnology makes everything by assembling it from the smallest, most basic building blocks. Other methods start with bulk matter and cut, grind, melt, and mold, or otherwise force it into useful forms.

This may not be immediately obvious, when you think of some of the earlier technologies we've considered. Sculpture, yes; more than one famous sculptor has observed that to sculpt a horse, you start with a block of marble and cut away everything that doesn't look like a horse. But what about a jet airplane or a skyscraper? Those are obviously bottom-up in the sense that they are very large structures assembled from smaller parts such as sheet-metal panels, rivets, girders, glass panes, wires, and switches. But those parts are themselves artificial, and making them requires cutting up, refining, and reshaping raw materials such as metal ore, crude oil, wood, and sand.

Making small things presents special challenges. A jeweler repairing a watch needs special tools—tiny screwdrivers, pliers, and such—for making and handling tiny parts, and magnifying lenses to see what he's doing with them. Those tools must necessarily be made with larger tools, and they themselves must be large enough to fit his hands. They're also subject to the limitation that even the steadiest hand is not perfectly steady, so they can't do things on a scale smaller than the smallest involuntary hand motions. There are ways around that—for example, building tools that scale the watchmaker's motions, acting as smaller and steadier hands—but there are clearly difficulties and limitations on how small people can make things by hand.

Richard Feynman, one of the most prominent physicists of the twentieth century (and incidentally an avid bongo drum player), in a 1959 lecture called "There's Plenty of Room at the Bottom," imagined a way to make objects too small to handle directly.[1] First you build a machine that can follow and imitate the operator's hands, but on a smaller scale. You use that small machine in a similar way to make a similar but still smaller machine.

And so on. But you can't keep doing it indefinitely. The smaller your work piece, the more important an error of any given size is; and there are always errors and "tolerances" in any mechanical work with bulk matter. Furthermore, when you get down to pieces small enough that individual molecules and atoms become significant features of the parts you're working on, you can't just cut arbitrarily smaller pieces. Doing so would cause fundamental changes in what you have, such as breaking molecules into smaller molecules of different chemical compounds. You'd have to contend with thermal and quantum effects that can be ignored at larger scales.

That's the province of nanotechnology: the range of sizes where the "landscape" is made of atoms and molecules, with sizes on the order of a few nanometers. The prefix *nano* means "billionth" (as the term is understood in America; an Englishman would say a thousand-millionth). A meter is a length that almost everybody in the world can visualize instantly, but most Americans have to be told is a bit more than a yard. So a nanometer is 1/1,000,000,000 of that, or 0.000000000025 inch.

And Feynman's method, while it makes an interesting thought experiment, is not the way to get into that realm.

STARTING AT THE BOTTOM

In the late 1970s, an MIT graduate student named K. Eric Drexler made a penetrating observation and an imaginative leap from what was known to what might be possible—a leap that may prove to be one of the most important and influential ever made.[2]

Drexler's key observation was that much of what happens in biology at the most basic level—the building of proteins, cells, and ultimately whole complex organisms from instructions coded in DNA, for example —was essentially similar to what's done in factories, but at a much smaller scale. The profound insight he added was that if organisms can carry out such nanotechnological processes with biological materials, humans might be able to do comparable ones with other materials. If true, such a capability might open up a mind-boggling range of possibilities that could transform human life, and the planet itself, more profoundly than even the control of fire, agriculture, the industrial revolution, or the information revolution we are now experiencing.

The ultimate goal and logical conclusion of Drexler's vision is a "molecular assembler": a nanoscale machine that can build something by using reactive molecules as "tools" to bond a few atoms at a time to the surface of a workpiece. This process could build up molecules or crystals, arranged as needed to form anything from a grain of salt to a skyscraper. Obviously *one* nanoscale machine couldn't build anything very big or complicated in any reasonable amount of time, so many of them would work in parallel, simultaneously building and connecting the parts of whatever larger structure they're ultimately intended to produce.

People have been using similar processes for centuries, but only in biological contexts and for macroscopically simple structures. Wine, cheese, and yogurt are all made by using self-replicating hordes of single-celled organisms to do molecular "magic" producing substances valued by humans. A key word here is *substance*, as distinct from *structure*. Wine, cheese, and yogurt are produced in bulk, to be sold in bottles or hunks; the microorganisms that make them do not build houses or air-planes out of them.

However, a counterpart of that does happen in all plants and animals.

The genetic material from a pair of pecan trees or people directs nano-manufacturing processes that not only build cells, but also build different *kinds* of cells for different purposes, assemble them into organs like leaves and livers and brains (and stop building when each organ is complete), and arrange all those organs to work together as an integrated, self-sustaining system. If Drexler's molecular assemblers can be built, they should be able to do the same thing—and extend the process to working with other materials and building kinds of objects that have never existed.

Some of the potential capabilities of nanotechnology sound so much like magic that it's tempting to think of it as *being* magic, but it isn't. It's constrained by the same physical laws as everything else, and there are some kinds of things those laws *don't* allow such machines to do. They could not, for example, transmute elements (that does happen, but involves high-energy processes *inside* atoms), negate gravity, or allow us to see the future. They can't even build just any kind of structure you might like; they're limited to ones allowed by the laws of physics and chemistry.

And it's certainly not reasonable to expect a "universal assembler," a concept that has often been mistakenly attributed to Drexler—a popular misconception that has hindered acceptance of nanotechnology by some parts of the scientific community. Expecting a single kind of nano-machine to act as a "general" or "universal" assembler would be like expecting a carpenter to build a house using a single tool: just try to imagine all the things that tool must do! On the other hand, a computer can be programmed to do an extremely wide range of tasks, from regulating a skyscraper's heating and cooling system to evaluating physics data to composing music. And the kind of assembler Drexler envisioned would incorporate a computer.

But it would also incorporate manipulators, tools for doing physical things. These would likely be several kinds of specialized assemblers, distantly analogous to hammers, drills, screwdrivers, and saws.

OBJECTIONS AND OBSTACLES

Can a computer really be made small enough to serve as a programmable brain for a nanomachine? Drexler and others have shown in considerable detail that it can—or, more precisely, that a computer that small can exist, using molecular-scale switching elements. Showing that something can exist without violating the laws of nature, however, is not the same as showing that human engineers can build it. The problem of getting from here to there can be a formidable one, and indeed currently looks like the biggest practical problem in achieving the kind of nanotechnology that Drexler envisioned. An international group of researchers led by Robert Freitas and Ralph Merkle called the Nanofactory Collaboration has put together the first R&D road map for nanofactory development and is actively pursuing this goal.[3]

Momentarily sidestepping the question of how we could make such tiny computers or other tools, how can we even know they could exist? The answer involves yet another convergence of technologies: "big" computers, like the ones most of us now use, can simulate the behavior of tiny systems.

Nanomachines will be, despite their small size, exceedingly complicated—even more complicated than microprocessors, and we've already seen how complex those are. If you've ever studied even basic physics, you probably remember problems such as calculating where a steadily accelerating particle will be, and how fast it will be going, at a given time. Starting from Newton's simple laws of motion, you can calculate nice neat equations telling you those things—provided there are no complications such as friction, which there always are in the real world. You probably also remember that when you try to take such complications into account, the equations get much more difficult—often so difficult that they can't be solved analytically. Similarly, it's pretty easy to calculate general equations for how two isolated bodies move under each other's gravitational influence, but impossible if you add even one additional body to the system.

What you can do in any *particular* such case is plug in the numbers for positions, speeds, and forces at one instant, and calculate how they will have changed a very short time later. Then you do it again with the new conditions, and so on. Do it often enough, for short enough time intervals, and you get a very accurate description of how the system behaves. The smaller the time intervals you use, the more accurate your description—and the more calculation you have to do. This is exactly the sort of problem for which digital computers shine: brute-force number crunching while you wait. They have made such calculations practical. They're called *simulations* because they give a detailed description of how a real mechanical or electrical system would act.

Chemical configurations can be simulated in this way, using a computer to figure out what forces act between all the atoms in a particular combination. Such calculations enable researchers to determine which combinations and arrangements will be stable, and what their properties will be. Drexler, Ralph Merkle, Robert Freitas, and other researchers have already applied such simulations to a wide range of nanotechnological tools, including computer elements, wheels, bearings, shafts, gears, motors, pumps, and batteries. (See plate 8.) The results are frequently astonishing: not only can those devices exist and work, but they can work exceedingly well.

Part of the reason for this is the fact, mentioned earlier, that scaling things up or down is not as simple as it sounds. We saw that everyday objects can't be made arbitrarily large because the mass that has to be supported increases much faster than the strength of the load-bearing members. That "square-cube law" works in the other direction when you scale things down, but another factor comes into play when you get down to the nanoscale. Atoms and molecules become "bumps," but soft bumps; and the peculiar effects of quantum mechanics, which often run contrary to common sense, become significant. One of the results is that no individual nanomachine does a large amount of work, but the output of each could be so large for its size, and so many of them could be used together, that the total capacity would be far larger than we'd intuitively expect. Drexler and J. Storrs Hall give calculations, for example, for a hundred-kilowatt engine (suitable for a typical car) made up of nanoengines massing only about fifty grams (less than two ounces).[4] It also turns out that some kinds of nanodevices have to work on different principles than their macroscopic counterparts. Electromagnetic motors, familiar from dozens of household examples, won't work at that scale; but electrostatic motors, completely impractical on the household scale, work just fine.

And a great many nanoscale motors working together can have large-scale effects, such as transporting people and goods.

Some have doubted that nanoscale machines could work because quantum effects and thermal motion of the molecules involved would prevent the parts from interacting properly. The simplest answer to that is refutation by example: living systems—like you, your cat, or the potted palm in your living room—are actual, functioning examples of large-scale systems made up of nanosystems. The fact that they live proves that nanosystems can exist that work at least that well—and suggests the possibility of even better ones. Nature, after all, is constrained by the way evolution works; it cannot suddenly produce organisms that can work in ways unrelated to those that came before. Every evolutionary advance is a modification of an earlier organism; but if engineers can learn to create what they want in the same size range, they may be able to design from scratch things that could never have developed in nature.

GETTING THERE FROM HERE

It's one thing to say that those astounding little things could exist, and to simulate how they would act if they did; it's quite another to make them in reality. Has any progress been made toward actually building those nanodevices that have been so tantalizingly simulated?

Some progress has occurred, and you've probably noticed that the term *nanotechnology* has already come into widespread use.[5] That's partly a reflection of real progress in the direction I've indicated, but not as much as you might think. Part of it is simply an expansion of the term to cover a wider range of things. Nanotechnology, as most commonly used in regard to things now on the market, simply refers to extremely fine powders and ultrathin films. The name fits in the sense that the particle sizes or thicknesses are smallish numbers of nanometers. They're useful for purposes like making stain-resistant or radiation-blocking fabrics, or "self-cleaning" glass. But the processes for making them are more extensions of ordinary chemistry and colloidal science than nanotechnology in the bottom-up sense.

"Buckytubes" are nanoscale tubes that are essentially rolled-up sheets of graphite, a very strong, slippery form of carbon. Their peculiar name derives from a convoluted series of associations: they were developed in connection with "buckminsterfullerene," another form of carbon (rather recently discovered, even though it occurs naturally in coal soot) named for its resemblance to the geodesic domes invented by the American designer/engineer Buckminster Fuller. Buckytubes are very strong and very slippery. They vary in shape and size and their other properties, such as electrical conductivity, vary too. They, too, can be made right now, by methods more chemical than strictly nanotechnological. Those methods produce a mixture of buckytubes, but researchers can pick out the ones with properties they want to study.

Other techniques that get a little closer to what Drexler envisioned include nanolithography and molecular beam epitaxy. *Lithography* originally referred to a form of printing done by etching on stone, but the term is now applied to methods for engraving, with light or electrons, the wires and transistors on microchips used in computers. Similar methods can be used for making other very small parts, but not as small as full-fledged nanotechnology will ultimately require. Molecular beam epitaxy involves depositing extremely thin films on surfaces, as in the very compact lasers used in CD and DVD players.

As noted earlier, many of the nanotechnological processes that go on in cells depend on a kind of self-assembly based on parts fitting together like lock and key. In DNA replication, for example, all the building blocks are floating around in aqueous solution, and a single strand becomes a complete double helix of DNA when a building block of the right kind bumps into each half rung on the single strand and attaches itself. Conceivably some of the processes used in a mature generalized nanotechnology will work in a similar way, and in fact some experimenters (such as Nadrian Seeman at New York University) have used pieces of DNA to build new kinds of simple machines.[6] J. Storrs Hall described in principle a method (which may or may not have been put into practice by the time you read this) for making nanoscale wires and transistors by attaching their components chemically to DNA fragments and tricking them into assembling in such a way that they will also pull the electronic circuit together.[7]

That's still not quite what we're looking for; the goal is not something that depends on things randomly bumping into each other in just the right way, but rather deliberate assembly by purposefully using a tool to bring parts together in the right way. A significant step in that direction was *scanning tunneling microscopy* (STM), a technique for which Gerd Binnig and Heinrich Rohrer won a Nobel Prize in 1986. Before their work, the idea of taking a picture showing individual molecules and atoms, and the soft curves of their electron clouds, seemed to most like unachievable fantasy. STM is not photography; it doesn't use light waves, which are much too long to make sharp images of atoms. Instead, a computer-controlled needle with an extremely sharp point (the tip is a single atom) is guided over the surface to be mapped. The computer uses the needle's electrical interaction with the electrons in the surface to "draw" a detailed, three-dimensional picture of the surface.

Many variations on the STM theme have been developed. As with other scanning methods we've looked at earlier, the ability to look has been extended to the ability to touch, and related devices have been used to move individual atoms into specific positions. Probably the first example was the initials *IBM* spelled out by thirty-five xenon atoms.[8] (See plate 9.)

The first successful positioning of single atoms—one of the key goals of nanotechnology—was a dramatic demonstration that that could be done. But it's still far short of the ultimate goal. Though individual atoms were moved and placed, the equipment to do it was (except for the business end) big and cumbersome, and the product too simple to be useful. To get molecular assemblers, we need to be able to make machines that are themselves on a molecular scale. As J. Storrs Hall puts it, "Current practice involves factories that make nanoscale products. Mature nanotechnology will involve nanoscale factories: working machines, atom-for-atom precise, that make things."[9]

How achievable that will turn out to be remains to be seen. Some critics, like the late Richard E. Smalley, doubt that we'll ever reach that point (but then, Smalley's arguments, once the most widely heard, have been refuted both theoretically and experimentally). Certainly there are problems that will have to be solved before Drexler's assemblers can be built; but while some scientists doubt that they ever can be, others with just as impressive credentials (like Drexler, Merkle, Freitas, and Hall) think that they can—and recently progress has been accelerating in that direction.[10]

History is full of scientists saying that this or that is impossible—manned flight, space travel, tiny computers, or heart transplants, for example—and then being proved wrong. The problem is often that they have such an intimate knowledge of (and vested interest in) their own specialties that they're all too familiar with the reasons why the methods they know can't solve a particular problem. What they overlook is the importance of convergence. Other people, working in other fields, may find approaches that the first scientists would never have thought of on their own. When the two (or more) lines of work converge, possibilities emerge that workers familiar only with either field could never have conceived. We've seen this over and over in the examples throughout this book, and we can reasonably expect to see it even more in the future. As Arthur C. Clarke said in *Profiles of the Future*, "When a distinguished but elderly scientist states that something is possible, he is almost certainly right. When he states that something is impossible, he is very probably wrong."

So we can't *know*, at least now, that the most advanced forms of nanotechnology envisioned by Drexler and others will ever come to pass. But if they do, the ramifications for the future of human (and other) life will be so profound that we need to be thinking about them well in advance—and we may not have nearly as much time as you'd expect. That's why Drexler and his colleagues founded the Foresight Institute: to encourage not only nanotechnology research, but also serious consideration of how the possibilities it may open up would affect human life—and how society might deal with those possibilities.

Why should the changes be so sweeping? The answer to that lies in the ways nanotech will converge with practically everything else—medicine, manufacturing, computing, communication, economics, government, and warfare, to name just a few. And the changes will be huge, no matter how far nanotechnology gets.

From here on we will take a look at those convergences. First we will simply survey some of the possibilities; then we'll focus on the exhilarating potentials for improving human life; then on the equally sobering dangers; and finally on how we might steer a wise, or at least prudent, course into that future.

CHAPTER 10
METACONVERGENCES
When Big Streams Make Still Bigger Streams

In previous chapters we've seen how convergences of different lines of development have produced the major streams of progress in which we are now immersed: for example, modern medicine, advanced computing, multimode electronics and communication, new arts, biotechnology, cognitive science, and nanotechnology. Now let's take a look at some of the still more dramatic changes that may result when *those* currents converge.

Just how dramatic they will be depends to a considerable extent on how far the development of nanotechnology goes. That area, perhaps more than any other, holds the potential to interact so strongly with all others as to produce changes far beyond anything else in human history. That is, *if* it reaches anything like the level envisioned by its most optimistic advocates. Since there is some uncertainty and controversy about whether and when that will happen, I will survey some of the possibilities in two categories: those whose possibilities can already be clearly foreseen, even without much additional development of nanotechnology, and those that would be opened up by a mature, full-fledged, fully utilized nanotechnology. Even the former include some startling possibilities, and the latter can be downright mind-boggling.

THE BARE MINIMUM

Even without new nanotechnology, we can already see without difficulty that today's currents and their convergences will take us into intriguing new territory. Some of those advances are already happening, and others seem practically inevitable.

We have discussed, for example, the medical possibilities being opened up by the confluence of general medical knowledge with new imaging methods, new understanding of genetics, and the ability to store and manipulate huge amounts of data. New imaging methods, fiber optics, and remote control technologies have allowed direct examination of internal medical problems, and far less invasive procedures for correcting them. Perhaps the best known and most popular of those are various procedures using lasers to repair defects in the eye. Problems that once would have required lifelong wearing of corrective lenses—with a new prescription every few years—are now fixed in a single office visit. After that, the patient needs no external lenses.

Some medical problems are, at least in part, genetically caused. We've already looked at some of the potentials and problems raised by the ability to read and modify the human genome. Those range from preventing genetic defects (and deciding what should be considered a defect) to correcting them after birth. When surgical intervention is still required, several more possibilities are opened up by convergences of different kinds of technology, including information and cognitive science. Those encompass *virtual reality*, *telepresence*, and *robotics*, all of which are closely related and so versatile that their applications extend far beyond medicine.

The term virtual reality (VR), like nanotechnology, has already come into widespread use to describe a considerably broader range of things than those for which it was originally coined. In its strictest (and most ambitious) sense, it refers to artificially generated sensory stimuli so complete and convincing that the person experiencing them can't distinguish them from reality. In a virtual environment, in the strictest sense of the term, a computer provides stimuli to all of a person's senses—sight, sound, smell, taste, touch, temperature, and so on—such that the person feels exactly as if he or she were in a situation very different from the actual one. A man might, for example, actually be lying in a bed in a darkened anechoic chamber in Boston, yet feel exactly as if he is skydiving into a Kentucky cornfield on a hot August afternoon, with a stiff wind, ever-changing cumulus clouds, birds singing all around, cattle lowing in a nearby barn, and the air full of the smell of freshly spread fertilizer. He feels the snap of his parachute opening and slowing his fall, and after he lands, he can walk across the field and look from side to side, or up at the sky or down at his feet, just as if the field and sky were real. No matter where he looks, there's as much detail as he takes the trouble to examine, and he can look at anything from any angle.

In short, the user has no way to tell whether the field, sky, and experience are physically real, or a computer-generated simulation.

In practice, nobody can come very close to that yet. "Virtual reality" games so far include highly realistic visual animation, with provision for shifting the apparent viewpoint in response to the player's head and body motions; high-quality stereo sound, with similar accommodations; and perhaps a joystick and pressure-sensitive glove that let a player steer a fictitious vehicle and fire weapons against phantom opponents, and sometimes feel localized resistance. But simulations of tactile sensation and smell are, at best, primitive and unconvincing. In many cases even the visuals don't hold up under really close inspection. Such a game may provide an exhilarating or scary simulation of a harrowing ride or battle, but few players are likely to forget where they really are.

But that possibility is coming. One current limitation is processing power: it takes a *lot* of computing speed and memory to handle enough data, fast enough, for a really believable simulation of even sight and sound. But remember Moore's Law; however much computing power you want (perhaps up to some limit that we're not even close to yet), it will probably be available if you wait a little while.

A more serious problem may be the human-machine interface. Earlier I mentioned the dramatic changes brought about by the introduction of graphical computer interfaces like those used by Macintosh and Windows operating systems. But thoroughly convincing artificial reality requires much more. At present, the sensory inputs in "virtual reality" come through the usual senses: visual imagery through eyes and audio through ears, via screens and speakers in a headset; tactile (if any) through gloves with transducers mounted in them. We can easily imagine tactile sensations being extended to more of the body by expanding the glove to a skintight suit, but that sort of thing is cumbersome, inconvenient to use, and probably uncomfortable. Further down the road, if cognitive science researchers learn enough about exactly how information is handled by the nervous system, it may become possible to bypass normal sensory inputs completely by coupling the VR hardware directly to the nervous system. Instead of projecting a picture in front of each eye, a future VR setup may feed directly to the visual part of the brain the kind of signal that it would get from the optic nerve in response to such a picture. It might even be possible to do that for other senses, such as smell, that don't lend themselves well to external simulations like pictures. Robert Freitas has described nanorobots that could do this.

Learning to do direct stimulation of sensory centers in the brain may be even more complicated than it sounds. In addition to the obvious hardware problems of getting signals directly into the brain without excessive physical trauma, there's the software problem of converting one kind of signal into a very different kind. So far our electronic computers are digital and programmed algorithmically. That is, they follow a set of instructions that tell the computer exactly what to do under precisely defined conditions, and memories are stored instantly and accurately in very specific locations.

The nervous system, in contrast, uses neural-net processing. It's taught rather than programmed, and learning takes trial and error and practice. Memories are distributed and associative. That is, a memory is stored in the network as a whole rather than in a single well-defined location, and it's connected somehow to other memories (which is why we use mnemonics such as the jingle to remember how many days are in each month). Individual nervous systems aren't identical. Even if two of them were, they might learn (program themselves) differently in response to similar inputs.

For all those reasons, it will be quite a challenge to develop an interface that will let a computer stimulate a living nervous system in complex ways, and respond to feedback from it. It will be even more of a challenge to develop such an interface that will work in the same way with all users. That may not even be possible. My personal hunch is that such interfaces will require at least parts of the external equipment to make more use of neural-net architecture; and even then, each machine and its individual user will have to train each other to work smoothly together.

In the meantime, even the relatively simple forms of VR already available have shown themselves useful in a number of ways—and not just in games. Entertainment is fine, but some applications of VR are life-and-death serious. VR is the basis for greatly improved flight simulators for training pilots, and for heads-up displays allowing pilots flying real missions at night or under instrument conditions to "see" the terrain they're flying over as if it were a clear day.

And, coming back to medicine, have you ever thought, as you were about to be anesthetized for surgery, that somebody had to be the first patient on whom your surgeon did this procedure—and you hoped it wasn't you? Virtual reality offers surgeons-in-training the chance to practice their moves as much as necessary on "patients" who can't be hurt (or sue) *before* they have to perform them on real people who can.

And, at a lower level of training, I've also heard of "virtual dissections" taking the place of real ones in high school and college biology classes.

ACTION AT A DISTANCE

Surgeons can use virtual reality in another way, too, which will lead us into the areas of remote manipulation and telepresence. Virtual reality makes you feel as if you're in an illusory but believable world, which may be constructed either from images (e.g., photographs and sound recordings) of real places and events, or entirely fictitious and computer-generated. Telepresence is similar, but lets you feel as if you're in a real but different place—and, in its strongest form, enables you to take an active role in what happens there.

The video teleconferencing that has already become fairly common in some business situations can be viewed as a very primitive form of telepresence, or an early ancestor of the real thing. People "meet" with others in distant places by sitting in a room with television images of the other participants (each of whom is in a similar situation). It isn't really telepresence because, while everyone can watch everyone else for the subtle visual cues that give the subtext of speech, basically they just talk. No one can forget that the others are really somewhere else.

Virtual reality is another obvious ancestor of true telepresence. Whereas virtual reality can give the illusion of being in a place that doesn't exist, telepresence lets you experience, with all your senses, what's actually happening right now in another place that does exist. But it does even more: it lets you influence what's happening there. For that we need one more ingredient: remote manipulation.

I've already mentioned this briefly in connection with Richard Feynman's "Room at the Bottom" speech, in which he imagined building small machines that could mimic the operator's actions and be used to build still smaller machines, even down to scales at which human fingers would be far too clumsy. Such motion-mimicking machines don't have to be very small, of course; they could just as well be very large—or of ordinary human scale. Why would you want such a machine to do work on a scale that humans could easily do themselves, especially if a human has to go through the same motions anyway? One obvious reason is that the work has to be done under conditions dangerous for humans, such as at high temperatures or deep under the ocean. Such remote manipulators have already come into widespread use in such situations. Probably the first real ones were used for processing highly radioactive materials. I've seen them used by a technician going through the required motions on one side of a very thick window while watching mechanical hands reproduce his every move to carry out the same actions on a radioactive sample on the other side.

Such remote manipulators are commonly called *waldoes*, after the story *Waldo* in which Robert A. Heinlein (writing as Anson McDonald) first described them.[1] (See plate 10.) If you look closely at Hubert Rogers's cover for that issue of *Astounding Science-Fiction*, you'll see the operator completely dwarfed by the machine holding its arms and "hands" in the same position as his. In this situation the job requires forces and stroke lengths far beyond human capacity. Simply slaving the machine's motions to the man's provides an intuitively simple way (at least for the operator, if not the programmer!) of getting the machine to do what's needed.

Surgical applications are the opposite extreme. Delicate operations inside the body can be done by little "robots" that carry out on a very small scale mechanical actions as directed by the much larger surgeon, allowing for more precision and less trauma to the patient. The surgeon doesn't even have to be in the same room—or, for that matter, on the same continent. As long as there's a good connection via radio and electronics to carry the surgeon's instructions to the robot, and real-time pictures of what's happening in the patient back to the surgeon, the work can be done safely and confidently. Radio signals only take about a fortieth of a second to go five thousand miles, so the doctor sees the results of each action in plenty of time to plan the next move.

But what if the work has to be done much farther away, so that the time lag becomes a serious problem? One area in which fine control of remotely controlled machinery would be highly desirable is space exploration—and eventually, perhaps, industry. In the last few decades we've learned more about our home solar system than in all previous human history, thanks largely to unmanned probes. Some of those just make remote measurements from space, but others land and release robotic rovers designed to move around, collect samples, and run tests on them, mostly without direct human intervention. This requires them to have some ability to size up conditions where they are and make decisions about what to do next—for example, not to continue in this direction because that will lead to toppling off a cliff. But so far they've been fairly simple-minded creatures, and from time to time they get into situations that they can't handle on their own. What do their human masters do then, given that (a) they have a heavily vested interest in getting as much good information from the robot as they can, and (b) they're millions of miles away?

ACTION AT A *LONG* DISTANCE, ROBOTS, AND ARTIFICIAL INTELLIGENCE

The term *robot* has been used to mean a wide range of things. The Czech writer Karel Čapek coined it to mean machines in roughly human form and capable of doing by themselves work that would normally be done by humans, for his 1920 play *R.U.R. (Rossum's Universal Robots)*. This is still the sense of the word most generally used and understood, thanks to a great many science fiction stories and movies about such machines. The Russian-American writer Isaac Asimov carried the concept a step further in his novels *The Caves of Steel*[2] and *The Naked Sun*.[3] Here a major character is a robot so realistically resembling a human that the other characters (and reader) must ponder the question of whether such a being (called an *android*) *is* in some sense human.

Such robots may eventually exist, but humans are very complex beings and imitating such "simple" actions as walking across a room has proved to be a huge challenge. There have been many robots in our midst for some years now, but most of them have been far from humanoid. Basically the term just refers to any machine that can carry out complex tasks automatically. Most of them have been manufacturing machines that could be programmed—for example, by punched tape as in the Jacquard loom, or later various forms of electronic digital storage—to do sequences of operations like cutting and drilling machine parts and then assembling them into something more complex. Others might be vaguely humanoid in form and/or function, and simply carry out mechanical operations under direct supervision by a human operator, like the waldoes described above.

Recently some have been more autonomous, often thanks to neural nets that let them learn to do things such as navigate across a room and avoid or climb over obstacles. A well-known example is a vacuum cleaner that runs by itself, moving around to cover a whole room, and then parks itself on a recharging station. Robotic pets, from dogs to dinosaurs, have gained some popularity. A recent alumni magazine from my former graduate school described the full-sized car that a team of its engineers is building to compete in a race with others of its class—without drivers.

The self-driven car is, at this point, exceptional. Almost all so far are pretty limited in their abilities. I've watched the vacuum cleaner be completely stymied by an embossed pattern in the center of a rug, mistaking it for a wall and cleaning the same couple of square feet over and over until somebody picked it up and took it away.

This is clearly not the sort of problem you'd want that expensive planet rover mentioned above to have. You'd like it to be able to make as many decisions as possible for itself. If it gets into trouble on Mars, it will take a long time for a controller watching it on Earth to realize there's a problem, and just as long to get an instruction back to the robot to fix it. Mars is always *at least* forty-six million miles from Earth, and often much more. That means a radio signal always takes more than four minutes to go from Earth to Mars or vice versa. Imagine yourself as the "parent" of a Mars-roving robot, anxiously watching its first steps and hoping it does well, but standing by to help it if it encounters something it can't handle on its own.

Imagine the depth of your disappointment as you see it about to plunge over a cliff—and realize as you jab the STOP button that it's already been lying broken at the bottom of that cliff for five or ten minutes.

For such situations, you'd like your robot explorer to be able to make decisions at as close to a human level as possible. In other words, you'd like it to have *artificial intelligence* (AI): the ability to do human things such as making judgments from incomplete or approximate data, learning from experience, and making free associations and intuitive leaps. Can a machine do that? Again, we can answer with a proof by example similar to the one used for nanomachinery: yes, because we *are* machines that do those things. A more meaningful question might be: can we learn to make, from scratch, other machines that can do those things?

Some very bright engineers and scientists, such as Marvin Minsky, John McCarthy, J. Storrs Hall, and Hans Moravec, think we can.[4] It's just a matter of learning in enough detail how we do what we do, so we can imitate it with other parts and materials. In other words, it requires a convergence, maybe not too far beyond ones we've already experienced, of cognitive science, information technology, and the other technologies that have already brought us to where we are: on a curve that is still steeply rising.

So we need to at least be thinking about the possibility that, in the not-too-distant future, we will be able to build computers, and even mobile robots, that can assess situations, make decisions, and act at a level of sophistication comparable to our own—and maybe well beyond. The big question then is: do we really want to?

Certainly such a machine would be very helpful in the Mars rover scenario. Since you can't be there in person or give instructions fast enough to do any good, you'd like your manufactured representative to be able to recognize unanticipated problems and figure out how to do something about them. Many of us would see obvious advantages in having machines much closer to home that can do tasks that we would rather not, such as housecleaning and yard maintenance. At present, our choices about such jobs usually reduce to doing them ourselves, hiring other people whom we have to pay and whose rights and feelings we have to respect, or using a machine too simple-minded to do a good job. Given a machine with true intelligence and the dexterity to do a wide range of mechanical jobs, we could enjoy being liberated from tedious tasks. We could sit back and do what we want to while something else does the work. We wouldn't have to worry about paying it enough or hurting its feelings because it is, after all, just a machine. It would be like having slaves, but without the guilt.

Or would it?

If we manage to build a machine that has enough intelligence to do human-level problem solving, which includes looking at things and asking "How does this work?" and "What if . . . ?", isn't it likely at some point to become aware of its own existence? Might it then think about how potential scenarios would affect its own future well-being, and how that in turn might affect its ability to do whatever it has been programmed to consider its primary goals? We're getting into philosophical territory here, but the problem may well become practical. Is there some point beyond which a machine cannot reasonably be regarded as "just a machine," and we need ethical codes to regulate how it and we behave toward one another? At some point engineers may have to consider what kind of ethical precepts can be built into a robot, whether it should have anything analogous to human emotions, and exactly how any of those things can be implemented at the design level.

Such questions have been the stuff of much science fiction, of which perhaps the best known are the many robot stories of Isaac Asimov.[5] Asimov's robots are constrained by three laws, paraphrased for human consumption as:

1. A robot may not injure a human being, or, through inaction, allow a human being to come to harm.
2. A robot must obey the orders given it by human beings, except where such orders would conflict with the first law.
3. A robot must protect its own existence, as long as such protection does not conflict with the first or second laws.

In other words, it's a hierarchical system, in which the first priority in all robotic decision making is human welfare, second is satisfying human desires unless doing so would somehow harm humans, and third is self-preservation unless that would either hurt humans or conflict with following orders. Asimov and his editor, John W. Campbell, worked it out as a set of rules that could reasonably be used as a basis for designing robots that could be both useful and nonthreatening. In earlier stories robots were too often portrayed simply as human-made monsters that turned on their creators. Some robots of that kind still show up in stories (especially in the visual media) and in laymen's imaginations, but many other stories have gone to the opposite extreme and taken for granted that real robots would have Asimov's Three Laws at the core of their programming.

In fact, they may or may not. Some builders may choose not to incorporate them, for reasons of their own, even if they can. Others may want to, but find that it's not as easy as it sounds. Even if they do, there is still room for uncertainty and conflict. Humans keep hordes of lawyers in business, wrangling over exactly how the law applies to real situations not anticipated when it was written (remember the Napster dispute?). Robots designed to follow the Three Laws might easily find themselves in situations where it isn't clear how the laws apply.

What if, for example, a human child under a robot's supervision wants to play in the street and her mother wants her taught not to do so under any circumstances? Can the robot let her do so if the block is temporarily closed to traffic? That will cause no physical harm now, but may help form a bad habit with the potential for causing harm later (and clearly involves disobeying the mother now). Or must the robot insist that the child stay out of the street even though it's perfectly safe at the moment, which will surely make her angry and hurt her feelings? It's easy to imagine situations in which the interpretation and application of the laws are ambiguous, or in which the robot's perception of how they apply may be based on misinformation. Asimov explored several such cases in his stories, and despite the potential for ambiguity, never found a better basic code.[6]

Nevertheless, some researchers think that something like Asimov's laws can be built into artificial intelligences at such a fundamental level that they will retain their primary importance even as the AI learns and improves itself. J. Storrs Hall has explored such questions in much more detail in his book *Beyond AI.*[7] Personally, I hope he's right, but am not quite so convinced. Still, I'd welcome the chance to see AI developed and see how it plays out. It seems to me that if any being, natural or artificial, has enough intelligence to think about itself and its surroundings, it's likely to come to some surprising conclusions—perhaps even ones that require abandoning precepts it had always considered beyond question. (A human analog would be a person who, after decades of life experience, suddenly finds or abandons religion, or converts from one to another.) And the more intelligent the being, the less predictable. More than one human business executive has complained that trying to manage PhDs is like trying to herd cats, and not without reason. Highly intelligent, educated people can be hard to supervise—but they can also be exceptionally interesting and fun to be around. We may find that the same can be said of some AIs, if we can accept them on their own terms.

A BIT FURTHER OUT

There are a few odds and ends I haven't yet mentioned here, some of which will lead us quickly into advances that are not quite so near-term, yet require nothing terribly far beyond what we can easily anticipate. For example, I've already mentioned little robots that do minimally invasive surgery inside patients under remote control from a doctor. Might we eventually get the doctor out of that picture altogether? We've already seen successful attempts along those lines, though considerably less sophisticated in what they do and how autonomously they do it.

Pacemakers—implantable devices that deliver electrical impulses to the heart to keep it beating at a steady rate—have been with us for quite a while. The basic concept, of jolting the heart into action with an electrical signal, goes all the way back to the late nineteenth century. The first internal pacemakers were tried out in the 1950s. Like most prototypes, they were only slightly successful and didn't last very long, but they proved that the principle was sound. Experiments over the next couple of decades ironed out a lot of the problems. Since the 1970s implanted pacemakers have become so common that we routinely see warning signs around other equipment (like microwave ovens) that can interfere with them, and they've kept many patients alive for many years.

Which raises the question: What else can be implanted?

Some pacemakers now have, if not intelligence, at least programming that allows them to monitor the wearer's changing needs and modify their actions accordingly. Often that programming can be changed as needed, from outside, by a cardiologist. We can easily imagine future pacemakers making more sophisticated appraisals of patients' immediate needs and more complicated changes in what they do in response. Some already incorporate defibrillators that can not only regulate the heartbeat when things are going well, but automatically restart it in the event of cardiac arrest.

We will probably see more and more instances of patients wearing implants that monitor multiple measures of health, and either take corrective action as needed on their own, or send a warning signal to the patient's doctor when something gets dangerously out of its normal range. Wearable blood pressure monitors have been helpful in gaining a better understanding of how blood pressure can be used as a health indicator. Getting a meaningful reading has always been a bit tricky because single readings can be thrown way off by such factors as stress—including that caused by being under a doctor's scrutiny ("white-coat syndrome"). Monitors worn around the clock can give a much better indication than a couple of readings in a doctor's office of how much a particular patient's blood pressure varies, within what limits, and under what conditions.

Diabetics' blood sugar can be monitored continuously, and the patient and/or doctor alerted when something needs to be done. A cranial counterpart of a pacemaker can watch for signs that an epileptic seizure is about to start, and administer electrical signals to prevent it. As test equipment and microprocessors get smaller and more capable, we can imagine more and more medical variables being monitored and adjusted automatically. Actual visits to a doctor's office or clinic will be necessary only when the patient's implants send the doctor a warning that something is getting beyond their ability to control. The doctor would then call the patient in to have it checked.

The anti-epilepsy pacemaker, acting directly on the brain, may be just a hint of things to come. I mentioned earlier the possibility of creating more realistic virtual reality by direct stimulation of sensory centers in the brain. That's a sizable step beyond nipping a seizure in the bud; but if the cognitive scientists, neurologists, and computer people can bring their fields together well enough to solve the interfacing problems, the possibilities go way beyond improving VR games and training modules. It may become a relatively simple matter for a compact implant to serve as a built-in computer, combining the strengths of the two broad types of computing. It would give the wearer the ability to do human-style associating and dreaming *and* the powerful, accurate number crunching we now associate with digital computers. Such a computer might even be able to hook directly, and often wirelessly, into a more evolved version of the internet.

A person equipped with such an implant would have enormous advantages over someone without one. He or she would have instant access to essentially all the informational wealth accumulated by the whole species, and be able to think about it in a fully human way—both alone and in rapid consultation with just about anyone else. Many science fiction stories have explored both the benefits that might emerge from such a tool, and the social problems that might arise from its widespread use. Would a person who preferred not to be so permanently and completely connected be able to resist the pressures to do so? Or would everyone be forced to choose at the outset between total immersion and total isolation—the latter implying something much less than second-class status?

On a more utilitarian note, what happens when cognitive science meets computer science and manufacturing technology to make possible "superwaldoes"? The remote manipulators I've already described have worked by means of a linkage, at least partly mechanical but maybe also partly electronic, to make a machine—large, small, or human-sized— imitate the physical motions of a human operator. But those physical motions are themselves controlled by neural signals sent through the operator's nervous system. Signals from the brain to motor neurons tell muscles what to do; signals from sensory organs back to the brain tell it what has been done so far and what needs to be done next. Why not "cut out the middleman" and couple the operator's nervous system directly to the control circuits of the waldo? Then simply thinking about what he's doing, while watching the results on a heads-up VR display, might let him control either industrial-strength or microsurgical manipulators as directly and intuitively as using his own hands and feet.

A step beyond that is the *cyborg* (short for "cybernetic organism"), a being that is part animal (often but not necessarily human) and part machine. A classic example is the science fiction character who used to be a spaceship captain but whose body was damaged beyond repair and his undamaged brain wired into his ship. He or she in effect *is* the ship, with a human brain still thinking and feeling but with ship's systems instead of arms and legs as its appendages. That may seem extreme and, to some, unlikely, but it's easy to imagine less extreme examples that might occur much sooner and more often.

Any person whose body has been badly damaged or worn out might welcome the chance for a fresh start with consciousness and memories intact, but with limbs and sense organs that work and don't ache. We have already made significant steps in this direction: there are people living relatively normal lives with artificial hearts and prosthetic limbs that work almost like the real thing. Cochlear implants make hearing possible for people who have never experienced it, and experimental prototypes of artificial eyes have been tested. If we learn to make replacements for most human organs, how many of them have to be installed before we cross the line from "human with accessories" to "cyborg"? Is such a line even a meaningful concept, or doesn't it matter?

The ultimate such replacement would be the replacement of *every-thing*—body and mind—either in more or less humanoid form or entirely in cyberspace. Information is information: a recording of your favorite song sounds the same when it's played, whether it's stored as a vinyl LP, open-reel tape, audiocassette, CD, or MP3 file. If your identity is the sum of all your memories and the programming that makes your brain work the way it does, it may eventually become possible to copy all that into a sufficiently complicated computer. This possibility is sometimes seen as a form of immortality: the body dies, but the personality lives on. Depending on where the copy is stored, it might control a robot that looks something like you, or even (if we combine this ability with some advanced biotechnology) a clone grown specifically to take over when your original biological body is worn out. Such a clone, programmed with your memories and traits, would not only think and act like you, but also be genetically identical and capable of reproducing and continuing the family line, either by further cloning or in the traditional way.

But is it really *you*? Or have you died and been replaced?

The answer depends on whom you ask, and quite possibly on how the process is done. In an essay called "Immortality for Whom?" I've argued that the copy is *not* you, even though it thinks it is and anybody conversing or otherwise interacting with it will be satisfied that it is.[8] Consider this analogy: If you use a computer program and set of data files stored on a disk (call it A), you can easily copy the entire contents onto another disk (call it B). Anything you could do with copy A, you can do exactly as well and in the same way with copy B. From your point of view, as the user, you have made the program and associated data "immortal." As long as at least one copy—it doesn't matter which one—exists, the data and program are safe.

Now suppose that the program you're copying is self-aware. It knows that it is A, and it has a deep-seated desire to protect its own existence. The copy also "knows" that it is A, so its need for continued existence is satisfied. But if the original A is erased or otherwise destroyed, it is gone, and the fact that B is carrying on (and still thinking it is A) does the original no good. The original A is gone.

Now substitute your name for A, and you may begin to have second thoughts about becoming immortal by having yourself "uploaded."

On the other hand, Hans Moravec, in *Mind Children*,[9] proposed an alternate version of the transfer scheme in which your mental contents are transferred a bit at a time and can be transferred back, with you remaining conscious through the whole process. In essence your mind is replaced bit by bit (a little like replacing one part of a car, and then another, and so on until the whole thing is new), and eventually it has all new hardware but continuous memory of being you. Something analogous happens with our bodies, with all the cells eventually dying off and being replaced with new ones, so that although you think of yourself as having been the same person since birth, you actually now have little if any original equipment left. It used to be thought that nerve cells were an exception, not being replaced or renewed like other cells, but recent research has found that this isn't nearly as true as we once thought.

If I were sure that the process would work as in Moravec's vision, I might well be interested. But we certainly can't do it now. We can't do the other method now, either—the kind that would destroy the original in the process of making the copy. However, it wouldn't require as big a leap. For the foreseeable future, I personally would want to give the matter a lot more thought before agreeing to have my mind destructively scanned so that somebody else could go on thinking he was me.

Quite likely not everyone will see it that way. People's philosophical guesses about what will happen will differ; if and when any kind of uploading becomes possible, some will want to take the plunge. Some may plunge even further: instead of transplanting their informational likeness into a robot or a cloned body, they may simply leave it in an interactive virtual community, or even a virtual world, that exists only inside a huge computer and is populated by many such uploaded personalities. Given enough computational ability and hardware, such a world could be so complex and detailed that the electronic personalities "living" in it would think of it as a real world as complete and satisfying as our own, and feel no need to venture out. If they *wanted* to interact with the physical world, they might be able to send copies of themselves forth in robotic bodies. But some might not bother; they might actually find their artificial world more satisfying than the "real" one, and be content to remain there indefinitely. Some thoroughly engaging science fiction stories have been set entirely in such virtual worlds, with no flesh-and-blood characters at all.[10]

That could be a problem, of course: if too many people retreat into cyberspace permanently, who will keep things running in the real world? The answer may be automated systems and artificial intelligence, though those too have their dangers. It's not a problem we'll have to face very soon, because both the scanning methods and computers of the magnitude required will almost certainly require significant input from the big stream I've been temporarily ignoring.

NOW ADD NANOTECH

The prospect of nanotechnology—of being able to manipulate and shape matter atom by atom—is impressive enough in itself. But when you combine that potential with the other currents we've been considering, everything changes dramatically. The magnitude and range of possible transformations is so great that I can only hint at it here. (For a more comprehensive survey and more detailed description of some of the possibilities, see *Nanofuture* by J. Storrs Hall.)[11]

I mentioned, for example, that advanced nanotechnology would probably be necessary to make personality uploads and large-scale virtual realities possible. That's an illustration of at least two broad areas where applications of nanotechnology could lead to huge increases in human capabilities: medicine and manufacturing. In medicine, scanning the information stored in the nervous system and translating it to a form that can be stored in its new "home" will require equipment that can get inside the entire nervous system and examine it in detail. In manufacturing, building a computer with enough memory and processing capacity to support a virtual world, in a reasonable space, will have to be done with nanotech methods. Many people may have little interest in uploading themselves or retiring to a virtual community; but nanomedicine and nanomanufacturing, if and when they become practical, will be used in a great many other ways as well.

Nanotechnology is a natural fit for medicine in several ways. Cancer, for example, is a collective name for a large number of diseases, all involving abnormal cell growth. Several kinds of nanotechnological remedies are possible, all potentially faster and more effective than any-thing we have now. Instead of making a large incision to remove malig-nant tissue, and hoping it hasn't metastasized to distant parts of the body, a future oncologist might simply inject a swarm of nanomachines into the appropriate body part to destroy, or perhaps even repair, the offending cells. Some of them might even circulate through the body, seeking cells of the cancerous type and destroying them wherever they're found.

Other nanomachines (or nanorobots), used in a similar way, might rebuild bones damaged by fractures or osteoporosis; or track down and destroy particular types of pathogens (disease-causing bacteria, fungi, and viruses).[12] Since aging itself is a surprisingly small set of problems with cell functioning, growth, and reproduction, cell-repair nanotech-nology has the potential to stop it in its tracks—and perhaps even to reverse it. Robert A. Freitas Jr. (whose book *Nanomedicine* is an excel-lent source for a more detailed exploration of such possibilities)[13] has described in some detail a process by which age-related deterioration might be not only stopped, but also reversed. The body would be restored piece by piece and in toto to the robust health and physiology of what-ever earlier age a patient might prefer.[14] Robert J. Sawyer, in his novel *Rollback*, has explored in depth some of the impacts such a process might have on the lives of some of the first people affected by it (which are not nearly as simple as you might guess).[15]

Surgical nanorobots might be particularly useful for some kinds of surgery, such as reattaching a severed hand and restoring full normal function, which involve making huge numbers of very tiny, delicate con-nections of capillaries and nerves. In a somewhat related vein (no pun intended), nanotechnology might be used to build in situ replacements for failing organs. Such replacements would fully take over the function of the natural organs and yet be clearly artificial constructs. They might thereby avoid at least some of the ethical controversies surrounding cloning and embryonic stem cells.

In our present world, replacement organs and prostheses are generally regarded as imperfect substitutes for the real thing. They're intended to restore some semblance of normal function, with no expectation of actually matching it. In a future with well-developed biotech and/or nanotech, this situation may change drastically. We may have the option of designing and making organs—including single-cell organs like blood cells—that work *better* than their natural counterparts. Robert A. Freitas has proposed what he calls a "respirocyte," that would store and transport oxygen in the blood much more efficiently than natural red cells.[16] Those could enable a person with severe respiratory problems to lead a normal life—or a normal person to live a supernormal one, breathing comfortably at altitudes so high that an unaugmented person would need supplemental oxygen. Other artificial enhancements might allow other improvements on natural human abilities, such as ultrasensitive sight or hearing.

As with nanotechnology in general, it seems at best a remote possibility that a single type of nanomachine could do everything that might ever be needed by any kind of cell or organ. But an arsenal (or "cocktail") of different types, each with its own skill set, could have much the same effect, and perhaps be almost as simple to use, at least at the doctor-patient level. What appears to be a single treatment might provide, for example, enhanced breathing, vision, and hearing.

Nanotech manufacture of substitute organs leads us to nanotech manufacture generally, which can be applied to almost (but not quite) anything. Remember that nanotechnology basically involves putting atoms and molecules together in desired configurations, so it can only work with the kinds of atoms that are available at the work site. So, for example, there's no way you can make a gold ring, no matter how sophisticated a nanotechnological synthesizer you have, if its stock of materials includes only carbon, hydrogen, oxygen, nitrogen, silicon, and sulfur. To make a gold ring, you have to have gold.

On the other hand, there's an awful lot that you could make with just the elements I've listed—which happen to be some of the ones that would be most abundant if you stoked your synthesizer's hopper with, say, kitchen garbage and soil from your yard.[17] Diamond, for instance, is hardly any problem, being just elemental carbon with the atoms put together in a particular way. Carbon comes in several structural forms. Some of them, such as diamond and buckytubes, are prime candidates for everyday building materials in a nanotech-based society. They're abundant, versatile, stronger than steel, lighter than aluminum, durable, and (with nanotech in widespread use) cheap.

Just what is this "synthesizer" I refer to? Nobody can describe one exactly, now. (If we could, we'd already be using them.) But in general it would be a physical setup containing a reservoir of material, a large number of programmable nanomachines to do various aspects of molecular-scale manufacture, and an interface for programming them. It might be quite large: for example, a "smart foundation" from which a skyscraper would be "grown," or a "factory garden" in which next year's crop of new cars would be planted and eventually harvested. Or it might be very small. Drexler, Hall, Freitas, and others foresee a society in which every home has a synthesizer: a tabletop appliance or piece of furniture, perhaps superficially looking a bit like a microwave oven or a refrigerator, used every day to make things such as food, dishes, and clothing.[18]

Many of those products may resemble their present-day counterparts, but their materials will be noteworthy. You might have diamond dishes, for example—to be thrown back into the synthesizer for recycling as raw materials when you want a new design or just don't want to clean them. Since objects will be made right at the point of use, with equipment that's extremely energy-efficient and not fussy about its feedstocks, they will be very cheap by our standards. (This may seem hard to believe at a time when energy costs are climbing rapidly, but bear in mind that oil refineries and gasoline engines are *very* fussy about what you put into them. Oil is an essentially nonrenewable resource, and much of what we now burn is used to transport raw materials to distant factories and manufactured products to distant consumers. And look at how the cost of electronics has plummeted and continues to plummet.) Recycling in general will become much more common, and storage less so. The prevalent attitude is likely to be, "Why store it when I can make a new one whenever I want it?"—and that won't reflect wastefulness, as it would now.

Food is a particularly interesting case. In our present world a great deal of energy is used to transport food long distances from farms to processors, from there to stores (wholesale, then retail), and from stores to homes or restaurants. Then additional energy is used to cook it, a process that involves not only heating it to a temperature that kills bacteria and pleases human palates, but also changing its chemical makeup (which is why cooked food tastes different than raw). Given sufficiently sophisticated nanotech synthesizers, almost all those steps can be eliminated. Feed in generic organic matter, program your menu, and collect the dishes and beverages of your choice—not in a raw state that requires further preparation, but cooked the way you like them and ready to go.

"Smart materials" can be made of what amount to vast numbers of nanorobots, capable of reconfiguring themselves quickly to meet changing needs. Windows, for example, can adjust their transparency to provide the indoor light level and hue that the occupants of a house want, day or night. They would compensate consistently for seasonal changes and shifting clouds—a little like the variable eyeglasses some people already wear, but much more so. A thin suit made of smart materials could constantly adjust how much heat it lets through—or generates, or pumps in the required direction—to keep the wearer comfortable under an extremely wide range of conditions. It could also change its appearance to suit (excuse the pun) the wearer's whims—not just from occasion to occasion, but continuously, displaying complex animated patterns or pictures if so desired.

Smart materials could be used anywhere. All the walls of a house, for example, could have the ability to display whatever "wallpaper," "paint job," or picture—still or moving—the occupants might want at a particular time. A part of one could serve as a mirror on request. Whole walls could glow uniformly, giving a more even light than any combination of lamps.

A particularly interesting and versatile example of smart matter is the "utility fog" invented (in principle, though not yet built) by J. Storrs Hall[19] (and described fully in the original article listed in the endnotes and more briefly in Hall's *Nanofuture* [pp. 188–95]). This is a "substance" made up of large numbers of microscopic robots, which, depending on their current programming, could float free like air molecules or assemble themselves into liquid or solid objects that look and feel like ordinary matter. They could assemble themselves into furniture, for example—or people. Let the utility fog concept converge with telepresence and you get the possibility of teleconferences that look and feel very much like "live" meetings, even unto participants shaking hands, offering each other drinks, and perhaps coming to blows if things went badly.

GETTING AROUND—AND BEYOND

Applying nanotechnology to basic fields like transportation offers a number of intriguing possibilities—especially in the area of aeronautics.

As we saw back in the chapter on flight, the ideal shape of an airplane differs quite a bit for different flying conditions. A plane well configured for high-altitude cruising in still air, for example, doesn't handle at all well at the crucial moment of landing in a turbulent crosswind, or vice versa. Taking off or landing on a very short field requires still other characteristics. The most extreme case is VTOL—vertical takeoff and landing —which requires a *very* different configuration. Airplanes so far have had to make lots of compromises; made mostly of rigid materials, they use arrangements like ailerons and flaps—pieces of otherwise stiff wings that can be temporarily moved into slightly different orientations—to approximate the kinds of performance needed under various conditions. The approximations are usually good enough to use, but often far from ideal.

A plane built with and incorporating advanced nanotechnology, on the other hand, could be equipped with lots of sensors to keep track of variables such as airspeed, temperature, pressure and wind. Onboard computers would calculate what hull and wing shapes would work best under specific conditions. Smart structural materials would continually reshape the plane into *exactly* the best shape instead of a crude approximation of it. (See plate 11.) Such technology might even be able to scratch one of the oldest itches in wishful futurology.

One of the favorite questions asked by people who like to scoff at the "predictive" abilities of science fiction (not that science fiction has ever claimed to be predictive) is, "Where is my flying car?" Many science fiction stories, serious and otherwise, have depicted futures in which ordinary citizens routinely get around in small aircraft, flying directly to and from homes, and often capable of doubling as ordinary cars on the ground. We now live in a world in which many science fictional visions have become commonplace, but that one (except for a very few extremely expensive experimental models that the FAA would never license in significant numbers) still seems stuck in fiction. Why?

Part of the reason lies in the requirements for the vehicle itself, and part in those for the operator and traffic control. The vehicle itself has to be an extreme example of reconfigurability: on the ground it must function as a car and in the air as a plane. The former *can't* have wings because they would constantly be bumping into things; the latter *must* have wings to be able to fly. So you need a way to extend wings for flight, and a way to fold or otherwise get them out of the way on the ground. The transition from ground travel to flight means taking off and landing, and if you want to be able to do that on ordinary driveways instead of airports with runways at least half a mile long, you'll need VTOL—which means you need a really extreme ability to reconfigure. You'll also need to do it quietly, if you're going to do it in residential neighborhoods.

Once you're up and away, who controls this thing? If many people are using them, what keeps them from running into one another? It's much harder to fly an airplane than to drive a car. I sometimes tell nonpilots who wonder what it's like that landing a light plane is a little like parallel-parking a car, except that you do it in three dimensions, at 80 miles per hour, with no reverse. Call me cynical, but I don't think you're going to get enough of the populace doing that well to turn them loose in the sky in anything like the numbers we now have on the road.

Unless they have help—and a suitable convergence of several technologies may provide enough help to make it a realistic vision. In *Nanofuture*, J. Storrs Hall describes in considerable detail a scheme that might make it work.[20] Highly efficient car-planes using nanotechnology, sensors, and computers conversant with aerodynamics would reconfigure themselves optimally for every phase of takeoff, flight, and landing. An autopilot using GPS and radio contact with a computer-automated traffic control system would make sure all the vehicles in the air get where they're going without hurting themselves or others.

Another effect of nanotechnology on transportation generally—not just aviation or any other specific form of it—is that much less of it will be needed. When people want or need to go somewhere, it will be easier; but it won't be *necessary* as often as it is now. With synthesizers making most things at the point of use from whatever happens to be handy, there will be far less need to transport materials and goods. In recent years we've already seen a dramatic increase in telecommuting. People go to a distant workplace less often and do more of their work at home, interacting with the outside world as necessary by telephone and internet. This trend is likely to continue and grow, especially as telepresence becomes better and cheaper so there's less and less perceived difference between teleconferencing and actual face time.

The same effect may spill over into private life, with more and more social visiting being done electronically rather than face-to-face. There will be ever less need to concentrate people in cities, for either business or personal reasons, especially with things like synthesizers, skin suits, and respirocytes enabling people to live in a wider range of environments, and with less dependence on near neighbors. Clifford D. Simak, as far back as the 1940s, explored how such decentralization might change human life and civilization, in his ironically named *City*.[21] Isaac Asimov showed an extreme development of this trend in his novel *The Naked Sun*,[22] where the inhabitants of Solaria normally interact in real time only by "viewing," and have a morbid dread of actually being in the same room with another person.

The trend toward decentralizing and expanding into previously uninhabitable environments can go beyond Earth. Colonies can and likely will be built on at least a few other natural bodies in our solar system, such as our Moon and Mars. In the late 1970s Gerard O'Neill, a physicist at Princeton University, developed the idea that other dwellings in space could be more easily built even closer to home: completely artificial habitats, which came to be known as "O'Neill colonies," in orbit around the Earth.[23] Those were usually envisioned as cylinders, rotating to provide a simulation of gravity, and big enough to house hundreds or thousands of people in an artificial ecosystem with parts resembling both cities and parks. (See plate 12.) Some of the material to build them could be brought up from Earth, and no doubt some of it would be—at first. But that's a very expensive way to do it. Once we had enough human presence in space, it's likely that most of the material for additional habitats would come from out there: the Moon and asteroids.

So far nothing has come of the idea. The initial expense of getting people and material into orbit from the bottom of Earth's gravity well is formidable, and the political climate has not been hospitable to making the huge initial commitment of resources and effort that would be needed to get such an undertaking off the ground. The political climate may change for a variety of reasons, such as scary events on Earth making more people realize that it's dangerous to humanity as a species to have all of us concentrated in one place. Nanotechnology may dramatically lower the cost.

The main reason it's currently so expensive to go to orbit is that it's done mostly with rockets. They're made largely of metal, which is heavy. They're commonly built in stages, for efficiency; but the stages powering the initial parts of acceleration are simply jettisoned and discarded after use. Most of their fuel does nothing except lift other fuel. All of those things make the transport cost per pound very high. Also, so far, most of the expense of spaceflight has gone into development of equipment that has been used very few times.

Nanotechnology can help in several ways. It has the potential to make spacecraft lighter, cheaper, and able to go all the way to orbit with a single reusable stage. Even near-term nanotechnology can provide ways to make substantially improved fuels.[24] If enough people become interested in going to space, the price of doing so can come way down. As we have seen over and over in such fields as electronics, the first units of something new are very expensive, to recoup the initial development costs. Once that's done, it costs relatively little to make additional units—and the more can be sold, the more cheaply they can be sold.

Since most of the direct cost of space travel is getting out of Earth's gravity well and into orbit—once you're there, as author G. Harry Stine used to say, you're "halfway to anywhere."[25] Once there's a sizable human presence in space, people may find it worthwhile and economically feasible to build very large-scale facilities to make that initial step easier. Various writers have proposed extremely tall towers, or space elevators ("beanstalks") connecting a point on the equator to a spaceport in geosynchronous orbit, stationary twenty-two thousand miles above the base.[26] Huge orbiting spaceports could act as momentum banks, storing energy and momentum from braking incoming spacecraft and using it to give a boost to departing craft.[27] All of those schemes have the advantages of allowing space travel away from Earth to begin above most of the atmosphere and gravity. All require, or at least benefit greatly from, the ultrastrong materials and special manufacturing techniques that nanotechnology may make possible.

Nanotechnology can benefit space-based visions of the future in one more important complex of ways. Nanomedicine will likely be able to fix at least some of the medical problems that now plague people who spend long periods in space, such as bone density loss—and in the process eliminate the need to spin habitats to simulate gravity. It may also repair radiation damage in cells, so that habitats will require less shielding than in older designs. Synthesizers may solve problems of food supply and recycling with unprecedented simplicity. Considering those factors and others, J. Storrs Hall imagines the possibility, perhaps by the middle of this century, of many quite small O'Neill colonies: single-family houses in space, affordable by families of ordinary means.[28]

PUT THEM ALL TOGETHER . . .

I've been describing some of the quite radical transformations that can result from the convergence of the many technological currents already in motion, but I've by no means exhausted the possibilities. I also want to stress again that hardly anything that confers such capabilities is likely to be either an unmixed blessing or an unmixed curse. What matters is not the technological capability in itself, but what we choose to do with it—and what others choose to do with it. So the same advance may be seen as a blessing by some and a curse by others.

Let me close this chapter with one example of a scenario that illustrates just what a convergence of several very advanced technologies might lead to. It's a scenario that I used in my novel *Argonaut*,[29] but it came to me in a roundabout way. It started when, in my capacity as editor of *Analog*, I read one too many manuscripts in which a handful of explorers slip into orbit around a planet none of them has ever seen before and begin studying it. In just a few days, they learn more about it than all the human scientists in history have learned about Earth in all the time leading up to now. "How can this writer expect me to believe they could actually do that?" I thought with some annoyance.

But then, coming up with answers to questions like that is one of the things science fiction writers do for a living. Evidently my subconscious took it as a challenge, and within a few days it had come up with an answer.

The key is a combination of telepresence, massively parallel computing, and nanotechnology. Suppose you can park yourself in orbit around your planet and immediately seed it with nanomachines capable of using local materials and energy sources to rapidly make lots of copies of themselves, and then other things. They all get busy making lots of little mobile information-gathering robots—say, vaguely insectlike—and relay stations through which they can send whatever data they collect back to you in orbit. You can collect a great deal of information, from all over the planet, in short order. Obviously you can't personally sift through it all, organize it, and understand it. But if you have a whole lot of molecular-scale computing power, some of that can be devoted to rapidly sorting through the accumulating data, looking for patterns of interest—such as the structure of ecosystems, weather patterns, and any civilizations that might happen to be present. When it finds one, it alerts you that, "Maybe you should take a look at this." Given sufficient AI in the system, it could quickly construct a pretty good map of the big picture, and then let you (or your robotic agents) take as detailed a look as you might like at particular parts of it.

I'll cheerfully grant that the ability to do all that is way beyond our present technological capabilities, but it's not beyond our ability to imagine. Nobody right now can tell you in detail how to do it (e.g., I'm not sure you could get enough energy where it's needed fast enough), but nobody a few decades ago could have told you how to build the computer I'm using to write this book, either. If you consider all that I've said so far about how converging technologies push each other along, accelerating overall change to previously inconceivable levels—and what people like Vinge and Drexler and Hall have said along similar lines—what I'm describing may be not only possible, but a lot closer than you'd think.

Such an information-gathering ability would be an exhilarating, quite possibly addictive, thing to have at your disposal. It would be a terrifying, quite possibly lethal, thing to have at someone else's disposal, if you or your country were the subject of such surveillance.

That dichotomy gives a good idea of the kind of choices we're going to be facing in the next few decades. The kinds of convergences we've been looking at will be putting increasingly vast amounts of power into human hands, subject to comparably vast uses and abuses. Sometimes people involved in the detailed imagining of such possibilities get so enthusiastic about the positive potentials that they gloss over the negative, just assuming that the problems will be solved and everyone will be as eager for the changes as they are. Sometimes others, less familiar with the details, are so frightened of the potential problems that they shy away from considering the benefits to be gained.

Neither extreme alone can lead us to the kind of balance we need. Both may play useful roles by acting as checks on each other, but our best chance for a future we like requires as many of us as possible to have as good an understanding as we can of both the potential rewards and the problems. What I'd like to do in my final chapters is briefly review first some of the gains those convergences may offer, then some of the dangers, and finally consider how we can weigh the two against each other and optimize our choices.

CHAPTER 11
POTENTIALS AND PROMISES

This chapter will be brief, not because the promises and potentials offered by the coming convergences are few or small, but because I've already said a great deal about them. However, so far I've focused primarily on specific possibilities, such as nanosurgery and single-family O'Neill colonies. In essence, we looked at the trees and not the forest. So at this point I'd like to back off and look at the big picture: broadly speaking, what can we gain from those potential advances?

Even without highly advanced nanotechnology, we can expect quite a bit. Some of the likeliest prospects are things that nearly everyone will agree are improvements, like better health, less traumatic treatment for medical problems that do arise, and longer lives.

Continued improvements in electronics, computing, and robotics, mixed in various combinations and tossed with better understanding of how we think and learn, promise more leisure time and/or the ability to pursue more interests. The things we'll be able to do include learning about new areas more easily and in greater depth, as well as new kinds of art, sports, and games. Games and education, by the way, are not as antithetical as many think. Most of the best teachers have long found that making learning fun (as long as that doesn't mean watering it down) makes it more effective, and clever programmers may increasingly blur the line that some think must exist between fun and learning.

More and more jobs that we'd rather not do can be handed over to machines. Some of those machines may be the humanoid robots of science fiction, but others may not even be obvious to the eye. Our houses themselves may do much of their own maintenance, automatically regulating temperature and humidity to our preferences, ordering supplies or repairs when they are needed, and reminding us of issues that need our personal attention.

Some possibilities may sound desirable to some but not to everyone. For the time being, however, I will mostly concentrate on what new abilities we might acquire, and postpone consideration of the controversies over how desirable they really are.

For example, human cloning, or even embryonic stem cell research, is seen by some as immoral, on religious or other grounds. But some will undoubtedly see it as a welcome alternative to traditional reproduction, for such reasons as difficulty conceiving or a personal preference for knowing what they're getting rather than playing a genetic lottery. Knowing what they're getting can go considerably further, from deliberately creating "designer children" at one extreme, to obtaining exact copies of an existing person at the other. The more drastic the change, the more controversy it's likely to engender. Many (but by no means all) would welcome the chance to ensure that their children won't have cystic fibrosis, hemophilia, or Huntington's chorea. Fewer (at least initially) would welcome such changes as creating people with gills instead of lungs so they could live freely and without heavy dependence on technological aids in the ocean.[1]

Back on less controversial ground, continuing improvement in the fusion of cognitive science with computer and communication technologies will give people easier access to an ever wider range of information. Implants smoothly interfaced to the brain may enable people to get the kind of information that they now get via search engines simply by thinking their questions, and they could do complicated calculations and correlations with the same kind of ease.

Governments will be able to provide stronger security at both individual and national levels by collecting and correlating data from many sources to detect terrorist plots and nip them in the bud. Some of this is already done, but it is likely to become more prevalent and more effective. For example, flight schools might routinely enter information about enrolling students in a database that flags student pilots who want to learn to take off but not land. Computers could search that information for connections to police records and identify individuals who seem likely to have terrorist ambitions. Pattern recognition software connected to security cameras in an airport might pick out a face in a crowd as belonging to such an individual, and prompt tight, targeted surveillance that could thwart a hijacking.

As transportation technologies continue to improve, and telecommuting and teleconferencing become more prevalent, travel and the pollution it produces may diminish, giving us a cleaner world. Moving some industry into space may help clean up Earth and slow resource depletion here by bringing materials from elsewhere—for example, the Moon and asteroids—and doing "dirty" manufacturing outside the atmosphere. It may even provide some additional living space, and new lifestyle options, in orbiting colonies and on the Moon and Mars. But how many of those space-based options we can do will be limited without help from nanotechnology.

A BIG BOOST FROM TINY TOOLS

If nanotechnology blossoms as envisioned by people like Drexler, Freitas, and Hall, all those tendencies will be made even stronger, by a dramatic margin. Not only will space colonization and industry become much easier and better able to make a large-scale contribution to human welfare, but practically everything right here on Earth will become easier and cheaper, which will open up a wide range of new options.

Health maintenance may become so good that practically anyone can be essentially immortal, or at least live and stay "youthful" for as many centuries as they care to. We're already at the frontier of routine telemonitoring of medical signs like pulse, blood pressure, and blood sugar, and noninvasive surgery when something needs fixing. With nano-technology, we may routinely ingest maintenance nanorobots that continually monitor everything and make adjustments at the molecular level, whenever they're needed, without our even being aware that anything is happening. Such maintenance may extend to repairing cancers and cardiovascular degeneration, even at stages too advanced for our present techniques to help.

When something truly drastic happens, like an accident causing massive trauma, repair may be effected not by hours of touch-and-go surgery, but by immersing the patient in a restorative brew populated by cell-repair nanorobots that rebuild everything as it was. Barring an accident so drastic that there's nothing left to repair, dying may become a matter of choice.

Even in the case of so terrible an accident, death may not have its present finality, if the essence of a person can be uploaded into a computer of the complexity made possible by nanotechnology. People might routinely store "backups" of themselves in this way, much as we now store backups of computer programs and files. Then, even if a person's biological body (or "meat suit," as Barry B. Longyear calls it in his Jaggers and Shad stories)[2] is utterly destroyed, the backup might simply be downloaded into a new one, allowing the person to live again—possibly in a new form.

The most obvious choice for a replacement body is a clone of the original, grown for the purpose and kept in unconscious stasis until needed as a spare. But that isn't the only option. A person might instead choose to be "reincarnated" in an entirely artificial body, perhaps made of new nanoengineered materials that give it superhuman strength and abilities. Or in a body of the opposite sex, or perhaps even a different species. That last would take a lot of getting used to, but advances in nanotechnology and cognitive science might help a human brain learn to control and use a new kind of body. Given the option, some might find it irresistibly tempting to experience life from a radically new point of view—and perhaps in a radically new environment.

Of which the most radical would be an entirely virtual one, which some might choose to use as an "electronic heaven." Whether such uploads and downloads are really the person from whom they're derived, or separate copies, is something we won't actually know until somebody does the experiment (and perhaps not even then). But if the experiment becomes possible, it's a fairly safe bet that somebody will try it.

Smart materials will carry the trend toward smart buildings, clothing, vehicles, and tools much further. Houses will be able not only to do some of their own housekeeping, but also quickly modify themselves in ways big and small, from changing the appearance of walls to changing the very locations and arrangements of walls. Furniture might be made to order on short notice, disposed of without fuss when no longer wanted, and remade in new forms when desired.[3] Food and clothing might be similarly made to order, used, and recycled unceremoniously. All of this would occur with far less energy use and pollution than our present methods. Such changes would transform our world to a cleaner one and our economy from one based on difficult distribution of scarce resources to one based on casual use (not consumption) of abundant ones.[4] With everybody, in effect, rich, many of the present causes of strife should evaporate, which one might reasonably hope would lead to a more peaceful world.

Smart, adaptable clothing and houses, capable of protecting their wearers or occupants and keeping them comfortable under extreme conditions, may enable people to live in many places now considered unsuitable, such as very high latitudes and altitudes and even underwater. This could relieve some of the population pressure we're feeling now, maybe even without overcrowding—at least for a while. Add communications networks that let people interact with little concern for where they are, and there will be fewer reasons for people to congregate in cities (unless they want to). They could, if they want, spread themselves more thinly over more places, including ones where hardly anyone now wants to live.

The same influences will make travel less often necessary, but easier for those who want to do it. Cars can be lighter, safer, more versatile, cheaper to own and operate, and automatic in every sense of the word. They may even become routinely airborne. If so, the whole system of roads may become less and less necessary, receive less and less use, and eventually be abandoned altogether (along with those nasty tolls!). This could be a real boost for the vision of a parklike planet. Not only are roads often unsightly and well known to disrupt ecosystems by fragmenting habitats, but they also may have other effects, less direct but no less important, such as influencing weather patterns. Extremely long continuous strips of asphalt or concrete heating in the sun may well affect the behavior of air above them. (On a recent summer drive I noticed that my car's outside-temperature gauge read several degrees higher on an expressway portion than on smaller roads nearby.)

And again, nanotechnology could do a lot for the exploration and exploitation of space. Without nanotech, the cost of getting into orbit and beyond is a major obstacle to large-scale industrialization, tourism, or colonization. With it, Hall's vision may become a real possibility, with so much capacity for expansion of population and industry that it might seem practically limitless (much as the American frontier once did).[5]

That would have the additional advantage of providing a sort of insurance policy for our species. As long as we're all in this one planet, a single planetwide disaster such as a major asteroid strike could wipe us all out in a remarkably short time. We know from the geological record that such catastrophes do occur, and why. The question is not whether it will ever happen again, but *when* the next one will be. It could be in a hundred million years—or it could be tomorrow. If we have any serious interest in ensuring the survival of our species and the worthwhile things it has created, we really need to get some of us planted elsewhere. With the capabilities now on the horizon, that can be not just a burdensome duty to posterity, but a way for many people to build rich and satisfying lives for themselves.

So: if we look only at the abilities we stand to gain from the coming convergences of medicine, information technology, communications, computing, robotics, cognitive science, biotechnology, and nanotechnology, the future looks dazzlingly bright. Long, healthy lives; the ability to be effectively wealthier and wiser than any of our ancestors; an unprecedentedly diverse smorgasbord of ways of life from which to choose; as much leisure and/or interesting activity as we want; a parklike planet finally at peace with itself . . .

It sounds almost too good to be true. Yet it's all quite possible, on the basis of science that we already know, at least in broad outlines. If we're prudent, careful, and lucky, we may actually be able to achieve something very much like it. But we can't take it for granted.

So what's the catch?

CHAPTER 12
PITFALLS AND PERILS

W e've just explored wonders that may advance the human race in a staggering array of ways. So what could be the downside?

The catch is, in simplest terms, twofold: (1) Any tool can be used in more than one way, and (2) Anything you do has unintended consequences—or, as John W. Campbell liked to say, "You can't do just one thing." In practice, there are additional complications, such as the one I hinted at earlier: not everyone agrees on what is good or desirable. Furthermore, some dangers are clear, obvious, and scary: the sorts of things that could destroy civilization, or at least kill large numbers of people. Others are subtle and insidious, but no less serious: not causing massive physical destruction, but "merely" big changes in the quality of life, likely through a series of steps so small that they go largely unnoticed until it's too late. In fact, the most obvious and scariest often turn out to be mostly imaginary, while the ones you're tempted to shrug off are the ones you really should have worried about.

GRAY GOO (ECOPHAGY)

One of the most hysteria-provoking ideas is so obvious that it became a popular catchphrase (first used by Drexler in *Engines of Creation*) soon after the public began to develop some vague awareness of nanotechnology. What if somebody develops nanomachines for some ostensibly benign purpose like cleaning up landfills or making furniture, and they get loose and start gobbling up everything in sight? In detail, that could mean a wide range of things, but the commonest version of the nightmare vision is that the runaway replicators break down everything organic they can find, like this year's corn crop, pets, you, and me. Other versions are equally imaginable: a proliferating swarm of nanothings that spread throughout the world and convert concrete or steel to dust would wreck the infrastructure of civilization—or the kind of synthesizer that can run on too many things, which is why I said earlier they were dangerous. Robert Freitas has even described "aerovores" that might wander the atmosphere, collecting carbon dioxide, extracting the carbon, and converting it to forms that block sunlight, possibly cooling the Earth's surface enough to plunge it into a new ice age.[1]

However, Freitas and Hall also point out the unlikeliness of any such scenario happening by an accidental release of industrial nanomachinery.[2] Nanomachines developed for manufacturing or medical purposes will be highly engineered and designed to operate very efficiently under very specific artificial conditions. Remove them from the special environment in which they're designed to function, and they'll simply quit. End of problem. (But we can't afford to be smug. Engineers can make mistakes, and it's difficult to foresee everything that could happen with very complex creations like computers or organisms. See, for example, Greg Bear's *Blood Music*[3] and Wil McCarthy's *Bloom*,[4] in both of which things do get out of hand and radically transform the world.)

A much more serious concern is the *deliberate* release of nanomachines that were designed to carry out such massively destructive actions. This concern applies not only to nanotechnology, but also at least equally (and perhaps even more) to biotechnology—because biology also involves replicators.

THE FERMI PLAGUE

The scariest scenario I've encountered in this vein is one that occurred to me several years ago in connection with one of several anthrax scares—and a conundrum first posed by the Italian physicist Enrico Fermi in 1943. It has bothered many scientists and writers ever since.

Even back then, some scientists recognized that there are many reasons to think that life is likely to arise as a natural extension of stellar and planetary evolution. So it should be fairly common in the universe at large. (Our recently gained ability to detect planets around other stars has only strengthened this impression.) Furthermore, while interstellar communication and especially travel are extremely difficult, they're not impossible—and there are a great many stars in our galaxy. Reasonable guesses at the relevant numbers suggest that we could reasonably expect to have encountered, or at least seen clear evidence of, at least one other technological civilization from somewhere outside this solar system.[5]

So why haven't we? Or, as Fermi succinctly put it, "Where are they?"

Many people have tried to come up with explanations.[6] They include civilizations failing to develop the necessary technology, destroying themselves before they reach a point where we would notice them, and avoiding contact with us because they're afraid of us or don't want to interfere with our developing in our own way. Maybe each of those scenarios, and the many others I haven't listed here, applies in one or more cases. But many people who've thought about the "Fermi paradox" think that if even one civilization reached a sufficiently advanced level, it would fill the galaxy in a relatively short time (by astronomical standards). Not everyone agrees (there's still considerable uncertainty and room for interpretation in the data on which such estimates are based),[7] but the possibility of extraterrestrial civilizations remains real. So, for the sake of argument, I tried to see if I could think of one thing that was likely to happen to most civilizations before they could go interstellar.

Unfortunately, I did. I say "unfortunately" because I don't like the answer—it's genuinely scary, and I find it all too believable. The essence of it is this:

1. As technology advances to the point where small groups or individuals can control very large amounts of power, it becomes possible for one malevolent or deranged person to do a great deal of damage—up to and including wiping out the population of a city, continent, or planet.
2. As population grows larger and larger, it becomes increasingly likely that such a person will arise.

Put those two observations together and you're forced to the conclusion that as the population and concentration of power grow, the probability becomes quite high that some nut with too much power at his disposal will destroy his civilization.

I first published that idea in an editorial called "The Fermi Plague" in October 1998.[8] I thought of it immediately on the morning of September 11, 2001, and posted it on the *Analog* Web site along with an introductory paragraph pointing out that what had just happened was a small-scale proof-of-principle of what I had said. Small-scale? Yes; the World Trade Center massacre "only" involved about three thousand people being killed by a handful of others, through an undesirable convergence of big-building and aeronautical technologies. But it was a pointed reminder that we've already reached a stage where individuals or small groups of them can do at least that much damage—or a lot more. A convergence of modern transportation, cities, genetic engineering, and biotechnology, for example, could quickly spread a highly lethal bioengineered plague around the planet. Nanotechnology isn't even needed.

We urgently need ways to make sure that doesn't happen—and we also need to make sure that whatever ways we come up with don't do too much collateral damage in other areas.

TWO-EDGED SWORDS

That's not as simple as it sounds, because the same tools can be used by either terrorists or governments trying to thwart them. They can also be used to turn governments *into* terrorists—or, as they're usually called in that context, tyrants, or totalitarian regimes.

We've already seen some of the tools that converging technologies have produced that can help thwart terrorists by identifying and monitoring them. Large databases, often conveniently linked to many other large databases and searchable by powerful computers, make it possible to collect a great deal of information about a great many people and sift through it looking for suspicious correlations. In a world without those databases and data-mining capabilities, no one in authority might ever notice that person A is requesting library books about explosives and exchanging e-mail with person B, who is exchanging e-mail with C, who is ordering some chemicals that could be used in making explosives from one company and other such chemicals from a different company. No one of those activities alone might attract interest, but the combination, detected by advanced technologies, might bring all three people under suspicion. That might lead to detecting and breaking up a terrorist plot.

Or it might wreck the lives and careers of three people who had completely innocent reasons for what they were doing and didn't even know about the others' actions. If you're a trusting sort, you might argue that if they're all innocent, that fact will come out and their names will be cleared. But history shows over and over that that doesn't always happen. Even if it does, it can take years—and even if the innocents caught up in the net are officially exonerated, they may never shed the stain on their reputations. Such things are especially likely to happen in a society recently jolted by a scare, such as a terrorist attack, into condoning emergency measures that it would normally consider unacceptable. So we need to be especially careful about striking a balance between allowing such intrusions for the sake of security, and limiting them to preserve the freedoms we're trying to protect.

Remember also that databases and data mining are by no means the only new double-edged tools we have that can be used for either protecting or undermining freedom. Where is all that data coming from? In addition to traditional sources, it can now come through such methods as tracing RFIDs embedded in vehicles, merchandise, passports, and even money that people use or handle—perhaps unknowingly. (It's easy, for instance, to attach a GPS transponder inconspicuously to someone else's car.) In recent years we've seen a strong tendency toward installing surveillance cameras in more and more places. Again, if you're trusting, you may argue that this will only hurt you if you're guilty of something and the photographic record provides evidence of that; if you're innocent, the same record can help clear you. But we've also seen a strong tendency toward most photography being digital, and digital photographs are so easily altered and manipulated that they've already lost almost all their value as evidence. Contrary to the quaint old adage, the camera *can* lie—so glibly and expertly that you can use it to prove pretty much whatever you want.

It boils down to this: governments now have an ever-growing array of powerful tools that can be used either to protect or to oppress their populations. In a democracy where the people have some control over their government, and exercise it diligently, we have a good chance to ensure that those tools are used mainly to protect. But no one must ever forget that any would-be Big Brother already has at his disposal means beyond any that George Orwell imagined.

THE MORE, THE MERRIER?

Two of the most attractive prospects held out by the coming convergences of technologies—especially if nanotechnology becomes a major part of the mix—are long, healthy lives and regaining the feeling that we can stop worrying about population because there are plenty of resources, living space, and room to grow. But it could be dangerous to become too attached to that view and stop thinking about population altogether.

Even if technology opens up lots of additional room for people on Earth, if those people keep reproducing freely without expanding beyond Earth, and people die less often, crowding will again build up. It will keep getting worse as long as population keeps growing. Even if nanotechnology and/or the movement of industry into space make the Earth a cleaner place, with parklike unpopulated areas and even some remnant wild spaces, there will be more people in line to enjoy those preserves.

We could easily wind up with parks and preserves that few people are ever able to enjoy or even visit. In that case, from the viewpoint of many humans, what's the point of having them?

If nanotechnology opens up space to colonization to the extent that some hope, that will provide some relief—but not an infinite amount. If we start filling near-Earth space with O'Neill colonies or something like them, the people living there will be reproducing and creating their own need for new ones. Some have suggested that they can build others while they're at it. As for Earth, some portion (between zero and all) of the population increase here can be dealt with by shuttling them off to live in space. This assumes it will be economically and otherwise practical to run a great deal of traffic between the surface and orbit, which may or may not be true. It also assumes that many people will *want* to move into space. This may well be true at certain periods of history, particularly if conditions on Earth are really unpleasant and many people see emigration to space as a brighter alternative. But then, that contradicts our assumption that converging technologies will have made Earth itself an inviting place.

However, tastes differ and fads come and go. It seems likely to me that some people will be eager to move into space and others will be reluctant, and the relative numbers of those types will fluctuate with time. During periods when life on Earth is good and few people want to leave, they will have to confront choices not all that different from those of the past or present. They can watch population grow, which will lead to what increasing numbers of them will see as deteriorating conditions. That may make some want to go, but if too few make that choice, others may be forced to leave (and who decides which ones born here will be forcibly deported, and who deals with the resulting resentments and hostilities?). And some may try to find ways to make people want to control their birth rates.

DEPENDENCE: THE PARKINSON PROBLEM AND ADDICTION

One cluster of potential problems, of which we are already well past the beginnings, has to do with dependence on technology. This has several intertwined aspects, some of which I've already touched on briefly. What

I called the "continuity problem," for example, is the danger of losing information about our past by storing it in a form that can only be accessed with the aid of a particular technology—and then losing that, either through disaster or (more likely) by switching to a new and incompatible technology. There are trade-offs. For example, stone tablets are far less dependent on supporting technologies, but they're also far less versatile and far less portable.

A second major "dependency trap" is what I sometimes call the "Parkinson problem." A twentieth-century truism, at least as true in the early twenty-first century, is widely known as Parkinson's Law: "Work expands to fill the time available for its completion." When British historian C. Northcote Parkinson formulated it in 1955,[9] it was primarily a comment on the human tendency to stretch jobs out as long as possible. Recently it has taken on a newer and arguably darker meaning.

Labor-saving devices reduce the time needed to do a particular task, which means that less time needs to be made available for that job. If that is done, more time is then available for other things. That *could* mean more time is then available to spend with family or friends, playing sports or engaging in hobbies, learning new things just for the fun of it, or doing whatever a person chooses to do. In practice, in our world, it often means instead that new jobs are invented to fill the time freed up by being able to do old ones faster. The result is what I call "the ironic epidemic": a population of people with more labor-saving devices than anyone before them, yet overworked, stressed-out, and unable to relax.[10] The epidemic is so bad, and so widespread, that it has spawned a new word in Japanese: *karoshi*, meaning "death by overwork."

It can get much worse if converging technologies continue to make it possible to work faster and faster, without people—individually and collectively—shifting their attitudes to compensate. I emphasize the latter, because technology in itself is not the problem; the problem is how people use it. Part of the blame lies with employers who think employees must be kept busy for a specified number of hours every week—or, increasingly, even more than the number they've agreed to. Technologies like e-mail, cell phones, and instant messaging make it possible for people to keep in touch all the time, from anywhere, so people feel that they (and others) must do so. Part of the blame lies with employees who go along with those demands, though they often don't feel that they have a choice. And part of it lies with a widespread tendency to jump on bandwagons and fully utilize whatever gimmick is currently being touted as the latest and greatest thing.

That can lead to a positive feedback analogous to its acoustic counterpart, the excruciating howl of a microphone and speaker placed too close together (which is caused by output from the speaker being fed back into the microphone and further amplified). Haste leads people to invent and adopt technologies that let them do things faster. The ability to work faster leads them to take on (or assign, or accept) more work, which leads them to find ways to work still faster . . .[11]

E-mail is a good example. The great thing about it is that it's so easy; the terrible thing about it is that it's so easy. It greatly facilitates many important communications. But it also encourages people to write and send vast amounts of drivel that they wouldn't bother with if it were just a little more trouble—and expect people to respond to it right away, simply because they can. (Very recently, though, I've started noticing a backlash: people largely or completely swearing off the internet, scrapping huge backlogs of unread and unanswered e-mail, and resolving to be very selective about what they read in the future.)

One more dependency trap might be viewed as the opposite of the overwork (or "haste makes haste") epidemic: a "dropping out" epidemic —a tendency for some people to become so wrapped up in artificial worlds that they effectively stop interacting with the "real" one. We've already seen early stages of this. Internet addiction and video game addiction have become recognized and named psychological disorders, with some therapists specializing in treating them. Why treat them? Because, as with any other addiction, people affected by them can become unable to function in their jobs and families, ultimately becoming burdens on those close to them or on society at large. In extreme (and, we can hope, unlikely) cases, such withdrawal into virtuality could become widespread enough to make a whole culture unable to sustain itself among others.[12]

THE DARK SIDE OF CLARKE'S LAW

While it doesn't seem likely that many whole cultures will get that wrapped up in video games or some future counterpart, there's a subtler trap that's more likely and, perhaps in the long run, a more serious danger.

Clarke's Law is another of those laws that is really an observation that was made by an astute individual who had one thing in mind, but which has recently acquired new shades of meaning. When Arthur (now Sir Arthur) C. Clarke first wrote, "Any sufficiently advanced technology is indistinguishable from magic,"[13] he was talking about the fact that the technology of one period (or a highly advanced alien civilization) would be so incomprehensible to people of an earlier or less advanced one that it would look like magic. That is how, for example, airplanes and flashlights must have appeared to members of isolated tribes having no prior contact with the outside world of the twentieth century. It's how my computer, car, and microwave oven would have looked to my ancestors in, say, the early nineteenth century; and how an ordinary household of a hundred years hence would look to most of us now.

But it occurred to me a few years ago that Clarke's Law now applies in a new way, not between cultures, but right here in our own culture.[14] To many people right now, our own technology has become indistinguishable from magic. They use cell phones, computers, digital cameras, MP3 players, microwave ovens, and cars with GPS systems with no understanding of how they work. As far as they are concerned, those things behave exactly like magic. Do the proper incantations and hand-waving and they do mysterious and useful things like talking to faraway people at will, showing movies or playing music on demand, and transporting people wherever they want to go, even if they don't know how to get there. What actually makes those things work is seen as the province of a special "priesthood" called scientists and engineers, and as being beyond either the need or ability of ordinary mortals to understand.

An important and dangerous result of this is that more and more people are simply using and depending on the fruits of science and technology, without understanding them (and often while deriding those who do as "nerds"). Meanwhile, fewer and fewer go into those professions. If that trend is real and continues, we could find ourselves increasingly dependent on very sophisticated technology that could stagnate and eventually decay because there aren't enough people left who understand it, to develop or even maintain it. Ironically, this problem could easily get worse as technology becomes more and more automatic and reliable. If it seldom fails, there will be less perceived need to know how to fix it. Rajnar Vajra vividly depicts an extreme case of this effect in his story "Emerald River, Pearl Sky," about a future in which technology is so automatic and so reliable that most people have literally forgotten that it *is* technology.[15] It's simply there, takes care of all ordinary needs without attention or thought on their part, and allows those with more than average understanding of what can be done with it to entertain and impress the others as "magicians."

But even they don't understand *how* it works—so when it starts to fail, who can stop the slide?

And systems *do* fail, eventually, with no known exceptions to date. A few years ago I experienced another small-scale but sobering demonstration when a localized glitch managed in a few seconds to shut down the electrical power grid over a large part of the northeastern United States and eastern Canada. Since most of the things people have come to regard as basic necessities—heating, air-conditioning, lighting, cooking, much transportation, computers, the internet, and so on—depend on that power, all those things ground to a halt, too. This was a big problem for everybody, and even more of one in places like Manhattan, where hundreds of thousands of commuters were trapped on a relatively small island with no way to get off.

This particular blackout lasted only a few days, and few lives were actually threatened by it, though millions were disrupted. But it could easily have lasted much longer, so the incident served as a good reminder of the vulnerabilities we impose on ourselves by allowing so much of our lives to become utterly dependent on a huge, massively interconnected infrastructure—and by concentrating vast numbers of ourselves in places where everything depends on that infrastructure.

During the long, exorbitant taxi ride home (I was one of the lucky ones who were able to get one), I found myself musing on whether we may have reached a point where we need to look for ways to make a major shift in the way we've been building our civilization.[16] How can we make ourselves less dependent on huge infrastructures that can be brought down so easily? Might cities themselves be an idea that has outlived its usefulness?

In the short term, making the big infrastructures more reliable and harder to disrupt is certainly an important goal. But is that just a stopgap, with no effect beyond increasing the mean time between major failures? We may be due for a more fundamental shift in our thinking.

I've already hinted at how future convergences of current technologies, especially if nanotechnology achieves its full potential, may provide such a fundamental shift. If people are able to make most of what they need right at home, with local materials and minimal energy usage, and if they are able to live comfortably in places now considered undesirable, those changes will dramatically reduce the need for both massively interconnected power systems and the concentration of people in cities.

Such dependence is by no means new. As James Burke wrote in 1978,[17] "The first man-made harvest freed mankind from total and passive dependence on the vagaries of nature, and at the same time tied him forever to the tools that set him free." This is true of every revolution through the most recent, so we are now very dependent indeed on a very complex web of tools. Those tools have brought us such enormous advantages that most of us would be very reluctant to give them up. Both the advantages and the dependence will be even greater with future technologies, so we need to be aware of the dependence and try to make it as little of a liability as possible. If we can do that, we should be able to gain advantages that will make the effort eminently worthwhile—but there's a good deal we must get through to reach them.

CHAPTER 13
GETTING THERE FROM HERE
Challenges and Strategies

We saw earlier that nanotechnology has the potential to make structures that haven't evolved naturally. Whereas new organisms result from an accumulation of single, relatively simple changes, nanotechnology could design completely new things from scratch. For example, it's easy to see how a leg could become a wing or a flipper by changing the relative sizes and means of attachment of the parts, but there's no simple way to change a leg into a wheel and axle. However, a mechanical engineer can figure out how a wheel and axle should work, make parts of the right size and shape from raw material, and put them together in the appropriate way. Nanoengineers may be able to do essentially the same thing, the main difference being that they would build their parts up from atoms and molecules while "traditional" engineers carve them out of bulk matter.

The problem that nanotechnology has now is figuring out how to get the process started. We could make nanotech assemblers if we already had nanotech assemblers, but at this point it looks as if we need assemblers to make assemblers. We know nanomachines could do all kinds of wonderful things, once we have a good set of tools for working at that level. But how do we get there from where we are now? As already mentioned, the Nanofactory Collaboration is hard at work on this problem.

We can draw considerable encouragement from the fact that the origin of life from nonlife involved a similar conundrum, and somehow it did happen. If I had to bet, I'd guess that we'll eventually find ways to make corresponding breakthroughs for other (nonbiological) forms of nanotechnology, too. But we'll still face a similar problem in getting from the kinds of society we now have to the kinds we could have as a result of converging technologies. A society based on cheap local synthesis of recyclable products by the people who will use them, who can live wherever they want, change sex at will, and live as long as they wish, will be so fundamentally different from the one we now have that it may not be easy to change gradually from one to the other. And we'll probably have to do it at least somewhat gradually, because we may not have the option of designing the new order from the ground up. Historically, "designed societies" don't have a good record.

The future can be unprecedentedly bright. But no matter how rewarding the destination, the way there will be challenging. And many of the challenges come not from technology per se, but from human nature.

PROBLEM OR OPPORTUNITY?

Some of the biggest and most immediate difficulties are likely to be economic. One of them I've already mentioned is "the ironic epidemic": the fact that the profusion of labor-saving devices in recent years has led not to a world of general leisure, but to one in which people keep getting busier and busier. When labor-saving devices make it possible to do jobs in less time, some workers are laid off while others stay on, working just as long as before (or even longer). Those who have lost their jobs agitate for the creation of new ones to provide employment for them, and society obliges by creating jobs—for their own sake, not because they need to be done. We often hear this or that program promoted as "creating jobs," as if jobs were an intrinsic good, a basic human necessity.

But they aren't. Basic human necessities are things like air, water, food, clothing, and shelter. Jobs are part of a complex social invention (actually less invented than evolved) for seeing that everybody gets those necessities. Nobody can have them unless somebody does the work of producing and distributing them. We employ people to do some of the work, and pay them with money that they can use to buy things that others produce. The system is a means to two ends: to enable individuals to get what they need, and to ensure that the society has enough of those things available for individuals.

The key word here is *means*. Jobs are a means to certain ends, not an end in themselves.[1] But our society is so geared to that particular means to those ends that priorities sometimes get confused. People take it as an unquestionable given that a certain amount of employment is necessary. They even define their own images of who they are largely in terms of what work they are paid to do. And when they're offered an unprecedented opportunity, they see it instead as a terrifying problem.

Consider a simple example: We sometimes hear arguments about whether it's better for the environment to take our groceries home in paper or plastic bags. There are trivial arguments that can be made either way, but the fact remains that both are disposable. A typical family will go through several thousand of them in a decade—when they could be doing the same job with five or six sturdy cloth bags that could last that entire time. The real choice is not between paper and plastic, but between disposable and reusable.

But if everybody switched to reusable bags, wouldn't that throw a lot of people out of work? People who make paper and plastic bags, for example; and those who collect and process petroleum and trees to make them, those who transport the raw and processed materials to where the bags will be made, those who transport the bags to the stores, and so on.

Well, it would certainly reduce the amount of all that work to be done. (It would also reduce the number of trees cut, the amount of irreplaceable petroleum used for making bags, the number of trucks needed, the amount of fuel burned, etc.) If we made comparable changes in many other possible areas, that would still further reduce the amount of work that needs to be done. Whether we see that as a bad thing depends on how stubbornly we fixate on one detail instead of considering the whole picture.

It's a problem if and only if we cling to the idea that everybody needs to be employed for forty hours every week. If you believe that, then if you reduce the amount of work by half, half your people are out of work and have trouble making ends meet. If, instead, you redistribute the work, everybody can work, everybody can make a living, and everybody only has to spend half as much time doing it. Everybody wins in a big way.

Even with the relatively modest labor-saving gains we've already made, we're already in a situation very much like that. We have an opportunity unprecedented in human history, to let everybody enjoy a high standard of living *and* have more free time to do things they'd like to do instead of things they have to do. Yet we persist in treating this opportunity as a problem. We struggle to "solve" it by inventing unnecessary work for people to do—which, in addition to wasting people's time, wastes huge amounts of energy and resources.

This can't go on indefinitely. We will have to find ways to convert our present system to one that treats the opportunity as what it is, better matching production of goods and use of people's time to real needs. It will involve a major reorientation of thinking and changing of ways. It can be done. Small steps toward it have been taken in the past, by such means as unions demanding and negotiating shorter work weeks. But we'll need to go much further in the future, even with the kinds of gains we're already quite sure will happen as the result of convergences like those of infotechnology and robotics. Given a full flowering of nanotechnology, the problem/opportunity will become much more pronounced. In the kind of world we can aspire to, everybody will have enough and nobody will have to work very hard to get it. (With a well-developed synthesizer technology, there may be very little that we *have* to do.) What we do with the liberated time will be up to us. It might well include *both* doing more work than we could before, including things we always wished we could but weren't able to, *and* having more leisure.

PREDATORS, PRIVACY, AND SECURITY

Many of our current conflicts stem from the fact that some people have a great deal and keep striving to get even more, while others have so little that they're chronically on the edge of starvation. If you think about how cranky you can get if you're just a couple of hours late for a meal, you can probably imagine, at least dimly, how billions of people, perpetually much hungrier than that, could account for a great deal of the world's strife.

On the flip side, you might imagine that having everybody well fed and physically comfortable might relieve much of that tension and allow at least most of us finally to coexist in peace. But would it really? Or are some people so geared to "clawing their way to the top" that they would keep right on doing it, even when there was no need to? If you read the news, you'll probably have no trouble naming people who seem to function that way—people who have no rationally believable need for more wealth, yet go right on ruthlessly doing whatever it takes to amass as much more as they can.

But then, given that there are such people, how much of their attitude is innate, and how much culturally conditioned? Our culture, at least, places a great deal of value on ambition and financial success, and that must have some effect on individuals who live in it—though it clearly doesn't affect everybody the same way.

It may not really matter how much is nature and how much nurture, since the way to any future world must begin with the present and the mix of people we now have, however they came to be the way they are. People now, in this part of the world, are conditioned to be highly competitive. When advances in biotechnology, infotechnology, nanotechnology, or any combination of them first become available, won't such people try to use them to gain unfair advantages before they become available to everybody?

Probably they will, and I'm not just referring to greedy business tycoons. It's equally likely that perfectly respectable corporations, interested in protecting their own interests and keeping their stockholders happy, will oppose anything that will eat into their established market shares. Big oil companies and electric utilities, for example, are not likely to be happy about technologies that would let individuals generate their own energy more cheaply than they could buy it. Unscrupulous and unaccountable governments might seize on surveillance techniques that would help them stamp out any opposition. On the other hand, governments that really want to protect their people's best interests could use those same techniques to nip terrorist plots in the bud. Terrorists, in turn, could use them to plan the most damaging attacks and carry them out.

Nobody will have a monopoly on those things; and as John W. Campbell liked to say, "Mother Nature is a blabbermouth. She'll tell anybody who asks the right questions." And, as he might now add, easy worldwide communications (especially the internet) make it hard to keep secrets. Converging technologies offer new tools to tyrants and terrorists, but also to those who oppose them. A modern counterpart of Orwell's Big Brother would find it a good deal harder to rewrite history convincingly, because it would be difficult (though not impossible) to wipe out all references to earlier versions, or keep them from being reinstated.[2]

Some of the most likely tools to be used in those ways are those involving the gathering, storage, and correlation of information: surveillance cameras (which can be widespread, mobile, and inconspicuous), RFIDs, massive databases, and data-mining software. Those devices are already in widespread use, so they're likely to become far more numerous, powerful, and pervasive. This trend has led to strong concerns about both privacy and security—and the questions are nowhere near as simple as some would make them sound.

On one hand, in a world where terrorists are active and converging technologies have given them the ability to do far more damage than their numbers might suggest, we need to protect ourselves from them. On the other hand, most of us are uncomfortable with the idea of every detail of our lives being open to scrutiny by anybody else, including our own government. (What if the government thinks *we* are terrorists, or pretends to for some other reason?) How can we strike a balance that will serve us well?

Some view privacy as paramount; others will cheerfully give up as much as they must to feel (but not necessarily be) secure. In practice, of course, we'll never have either—absolute security or absolute privacy—though we could drive ourselves and everyone around us crazy chasing the illusion of one or the other. Some maintain that so much of what we do is already recorded and tracked electronically that privacy is already dead. Some claim they don't care because they have nothing to hide—but it's presumptuous of them to assume that everyone else does or should feel the same way. It isn't necessarily too late to make some decisions about how the balance goes.

Some particularly thought-provoking ideas on those matters can be found in David Brin's book *The Transparent Society*[3] and in Robert J. Sawyer's Neanderthal Parallax trilogy of novels: *Hominids*,[4] *Humans*,[5] and *Hybrids*.[6] Brin is concerned about privacy, but also about the possibility that a one-sided overemphasis on that could actually reduce freedom by allowing even more skullduggery than we now have to be carried out in secret by parties of all sorts: corporate, criminal, and governmental. He suggests that it's already too late to avoid widespread surveillance, but maybe we can turn it to our advantage. Freedom (if not privacy) might be enhanced by making the information flow both ways, with citizens able to observe their government as easily as vice versa.

Sawyer's novels show an alternate history in which Neanderthals have developed an advanced civilization along quite different lines from (and acting as a mirror for) our own. There every citizen has a "Companion," an implant that makes a complete recording of the wearer's life and stores it in a central database (the "Alibi Archives"), not accessible to anyone else except under very special circumstances. The arrangement provides plenty of privacy under all normal circumstances, but keeps incontrovertible evidence available for any rare case in which it might be needed. Sawyer's Neanderthals are significantly different from us in fundamental character and attitudes, so what worked for them may not work for us. But then, human attitudes vary a lot more than most people realize from one culture to another, and almost everybody assumes that his particular culture's traits and values represent "human nature" in general.

HUMAN NATURE: CONSTANT OR VARIABLE?

There's a popular cliché that says, "You can't change human nature." But is it really true? Or might some of the advances we've been looking at change even that?

At some level and over some period of time, there probably are a few general characteristics that hold true for people of many different times and places. But there are also plenty of differences, and changes across time. Attitudes and values are certainly different among, say, the motorcycle culture in 2007 Los Angeles, contemporary Iran, an Amish community in nineteenth-century Pennsylvania, the court of Louis XVI in France, and an Australian Aboriginal tribe four hundred years ago. Most people in today's America would find the notion that it's acceptable for people to own others of another "race" quite foreign to their natures. Some of their ancestors two centuries ago would have found it equally hard to embrace the notion that it isn't.

So it is demonstrably possible for people to differ dramatically in their most deeply ingrained beliefs. But biotechnology, genetic engineering, and cognitive science may lead to much more dramatic changes. If humans can be given implants that put them in perpetual mind-to-mind contact with many others, bodies that can fly or live underwater, or robotic bodies, what does that do to their humanness? Certainly they will experience the world in significantly different ways than we do. How will that affect the way we should treat them, or what we can require of the way they treat us? What if you physically die, but have your consciousness copied into a computer where it can be expected to survive indefinitely? If you're a person I love or admire, I may welcome the chance to continue visiting and conversing with that version of you (let's call it you-2). But should you-2 be allowed to vote in elections about matters affecting only people with standard biological bodies? And should I, as one of those, be allowed to vote on matters affecting only people like you-2? Could a totalitarian regime exercise direct mind control over its citizens, making questions about voting moot?[7]

Those questions are neither simple nor frivolous, and we may need answers to them sooner than you'd guess. And those answers may require some radical rethinking of what we mean by *human*.

HOW MUCH OF THIS WILL HAPPEN, AND HOW MUCH CHOICE DO WE HAVE?

Not everything I've described in this book will actually happen—and many things that I haven't mentioned (including some that neither of us has even imagined) will. Prediction is a difficult and chancy business. Some things that we know *could* happen, don't, or take much longer than we'd expect, often for reasons having nothing to do with their technical feasibility. Others suddenly take off and grow beyond anybody's wildest expectations, often for reasons having to do with converging technologies. And occasionally there will be a complete surprise: a discovery from so far out in left field that nobody could have seen it coming, but that turns out to have profound effects on everything that follows.

A few examples:

I've mentioned holography as something that was first imagined at a time when the supporting technology didn't exist to make it reasonably easy. When that came along with the laser, holography became a hot field for research. I did some myself, so I was pretty familiar with what was being done. When I used a real-time holographic television system in my first published science fiction story,[8] I deliberately set it much closer (within thirty years) to the time when I was writing than anybody working in the field considered possible. Five years later I read a research paper that said casually that the equipment for real-time holographic television currently cost about thirty thousand dollars, but that would come down if anybody was sufficiently interested. Evidently nobody was, because I've hardly heard a word about holographic TV since then.

Cheap, reliable fusion power has been a couple of decades in the future for a lot longer than a couple of decades.

On the other hand, computers have become far smaller, more powerful, more ubiquitous, more interconnected, and more pervasive in their influence on society than almost anybody in science fiction ever imagined. Relativity and quantum mechanics, which would have sounded ridiculous to almost anybody 150 years ago, have radically reshaped the entire landscape of physics, and much of technology.

Science fiction does not try to predict the future, but only to imagine as many as possible paths that it *could* take. That is what all of us want: to prepare for whatever future we actually get. And it's what we *must* do, because we can't afford to be blindsided if we can possibly avoid it. The possible consequences are just too big. So we need to imagine everything we can that *could* happen, how it might impact our lives, and how we as individuals and societies could deal with it. Science fiction can help, but it's not enough. Science fictional imaginings are usually, by necessity, fairly qualitative. To think seriously about what we might be up against— and what we might be able to achieve—we need to also look at some of the more rigorous thinking being done (e.g., the quantitative calculations in the works I've referenced by K. Eric Drexler, Robert A. Freitas Jr., and J. Storrs Hall).

Some people find the thought of drastic change intimidating. They would rather try to avoid the whole thing by banning this or that— nanotechnology or stem cell research, for instance. But bans won't work. Not only does that approach amount to throwing out the baby with the bathwater, but also to passing up golden opportunities for fear that they're booby-trapped. Moreover, Mother Nature really is a blabber-mouth. If something can be done, somebody will do it, even if we don't— so we need to learn as much as we can about it, if only so we can defend ourselves against somebody else's abuses. As J. Storrs Hall points out,[9] some of the nasty uses of nanotechnology could be devastating if somebody used them on us at our present state of development—but a mere nuisance if we had our own nanotech defenses against them. So the safest world is likely to come not from bans and secrecy, but from open development in as many places as possible—not just for nanotech, but for any new technology.

We'll still have plenty of choices to make, both individually and collectively, and we'll need to make them with as much knowledge and understanding as we can muster. To do this, we'll need good, solid education at all levels, from the most basic to the most advanced.

Basic? I've had a pizza clerk say she couldn't give me change because the computer was down and she couldn't figure out how much she owed me without it. None of us should be *that* dependent on the tools that let us do harder things.

Advanced? Probably the biggest thing we need here is more interdisciplinary education, with more people in different fields learning to talk to one another. As we've seen, biotechnology and nanotechnology are themselves products of interdisciplinary convergences, and now they are converging with still other areas. Recently "complex systems," which can be anything from a living cell to an economy to a global atmosphere or ecosystem, has been emerging as a major field of study in its own right—perhaps the most interdisciplinary of all, and in the long run, perhaps the most significant.[10]

At a time when little is certain except the imminence of sweeping change, we are surrounded by voices warning us of imminent doom, and others promising a glowing future of nearly unlimited potential. Either of those extremes is possible. But it seems to me more likely that we'll get something in between—not perfect, not forever free of problems, but, if we play our cards right, maybe better than anyone before us has enjoyed.

To make that happen, we'll need determination and informed choices. To make those choices wisely, we'll need to understand the forces that will shape our future. Not just single currents like electronics, medicine, or space exploration, but the way all those currents converge and interact.

And we'll need to understand that we're all in this together—that none of us alone can choose what the future will be. On one level, some of our possible futures will offer us more options as individuals than we've ever had before. But on another level, some choices must, by their very nature, be made by all of us together—because what each of us decides will inevitably affect others. One person cannot decide, for example, to live on an uncrowded world, if everybody else decides to make a crowded one and keep no roads open to go elsewhere.

The stakes have never been higher. Let's try to do it right.

NOTES

INTRODUCTION. CONVERGING CURRENTS: THEN, NOW, AND TOMORROW

1. K. Eric Drexler, *Engines of Creation* (Garden City, NY: Anchor Press/ Doubleday, 1986), esp. pp. 217–30.

2. Emmett N. Leith and Juris Upatnieks, "Photography by Laser," *Scientific American*, June 1965, p. 24.

3. Vernor Vinge, *The Peace War* (New York: Bluejay Books, 1984).

4. Drexler, *Engines of Creation*, pp. 217–30.

5. Ted Nelson, *Literary Machines* (Swarthmore, PA: Ted Nelson, 1981). The word "hypertext" appears as early as the first page and in many other places; the book itself uses a hypertext-like organizational scheme.

6. Although Drexler was instrumental in founding the Foresight Institute, by 2007 he was no longer associated with it. However, it remains active as the Foresight Nanotech Institute. People interested in keeping up with progress in this field and helping to further its goals can become members and receive a regular e-newsletter through the institute's Web site, http://www.foresight/org.

7. Mihail C. Roco and William Sims Bainbridge, eds., *Converging Technologies for Improving Human Performance* (Arlington, VA: National Science Foundation, 2002).

8. Ibid., p. 1.

9. Stanley Schmidt, *Argonaut* (New York: Tor, 2002).

CHAPTER 1. FROM FABRIC LOOMS TO THE INTERNET: THE STORY OF COMPUTING

1. *Encyclopaedia Britannica*, 15th ed., 29 vols. (Chicago: Encyclopaedia Britannica, 1987), 6:467, 14:530.

2. Isaac Asimov, *Isaac Asimov's Biographical Encyclopedia of Science and Technology*, new rev. ed. (New York: Doubleday, 1972), pp. 288–89. See also *Encyclopaedia Britannica*, 1:765, 4:92, 16:680, 28:477.

Both the *Britannica* and *Asimov's Biographical Encyclopedia* are useful sources of historical and biographical information about technological development, particularly in its earlier phases. However, the lives of the people who created our technologies are as intertwined as the technologies themselves. When I refer to these sources I will typically list the *principal* references to the subject at hand; but the best way to get a feeling for how these fields developed is to curl up with the books, start reading where I suggest, and follow cross-references as far as you care to.

3. Asimov, *Biographical Encyclopedia*, pp. 647–50; *Encyclopaedia Britannica*, 4:370, 11:701, 17:1049.

4. Asimov, *Biographical Encyclopedia*, pp. 567–68; *Encyclopaedia Britannica*, 7:826, 15:230, 28:471.

5. Asimov, *Biographical Encyclopedia*, pp. 564–65; *Encyclopaedia Britannica*, 3:929, 18:314.

6. Asimov, *Biographical Encyclopedia*, pp. 297–300, 279–80; *Encyclopaedia Britannica*, 5:854, 28:486, 8:340, 28:471, 28:486.

7. Asimov, *Biographical Encyclopedia*, pp. 450–51; *Encyclopaedia Britannica*, 2:67, 28:471, 28:495.

8. Asimov, *Biographical Encyclopedia*, p. 450.

9. Leo Vernon, "Tools for Brains," *Astounding Science-Fiction* 23, no. 5 (July 1939): 122.

CHAPTER 2. AVIATION AND BIG BUILDINGS

1. Alan Howard Stratford, "Air Transportation," in *Encyclopaedia Britannica*, 15th ed. (Chicago: Encyclopaedia Britannica, 1987), 28:821.

2. Ibid., 28:825.

3. For more detail on how wing shape and orientation affect lift, see, for example, *Pilot's Handbook of Aeronautical Knowledge* (Washington, DC: US Department of Transportation, 1980), pp. 1–12.

4. Both "ailerons" and "flaps" are actually flaps, or portions of the wing that can be moved into different orientations to alter flight characteristics. Aircraft designers and pilots use different terms to specify the locations and functions of such movable flaps. Ailerons, for example, are used mainly in steering, while flaps are used to adjust lift for situations such as steep landing approaches or takeoffs.

5. *New York Times*, December 10, 1903. Further details on the original editorial seem to have been lost; but numerous other sources agree on the date and content, including Isaac Asimov, *Isaac Asimov's Biographical Encyclopedia of Science and Technology*, new rev. ed. (New York: Doubleday, 1972), p. 412.

6. When water is boiled, the steam produced has a temperature equal to the boiling point of water (212 degrees Fahrenheit or 100 Celsius). A superheater is a coil in which steam, after vaporization, is heated to even higher temperatures.

7. Asimov, *Biographical Encyclopedia*, p. 196.

8. A. I. Root, *Gleanings in Bee Culture* (January 1905): 36. Root's candle business is still operating in Medina, and the full text of the article can now be read at http://www.rootcandles.com/about/wrightbrothers.cfm (accessed August 16, 2007).

9. A good illustrated introduction to the history and problems of constructing large buildings can be found in David Macaulay, *Building Big* (New York: Houghton Mifflin, 2000), pp. 129–91.

10. David Brin, *The ArchiTechs* (television special, the History Channel, October 11, 2006).

11. However, we'll look later at some reasons why they may lose some of their present importance.

CHAPTER 3. NEW ARTS AND SCIENCES

1. "Sketches of Frank Gehry," in the series *American Masters*, first aired on PBS, September 27, 2006.

2. E.g., "Architect Frank Gehry Finds CAD a Boon to Art and Business," *CAD Digest*, http://www.caddigest.com/subjects/aec/select/022304_day_gehry .htm. For an introduction to CAD itself, start with http://en.wikipedia.org/wiki/ CAD (both accessed August 16, 2007).

3. "Moog Synthesizer," http://en.wikipedia.org/wiki/CAD/Moog_synthesizer (accessed August 16, 2007).

4. George Johnson, "Undiscovered Bach? No, a Computer Wrote It," *New York Times*, November 11, 1997; also available at http://query.nytimes.com/gst/ fullpage.html?res=9904E5DC1039F932A25752C1A961958260 (accessed August 16, 2007).

5. These techniques are also useful in the creation of still pictures, and many "painters" now work entirely with computers, not using canvas or paints at all. They can even develop a scene in three dimensions, so that if they submit one version and an editor would prefer it from another viewpoint, the computer can change it quickly and easily.

6. Grey Rollins, "The Ghost in the Machine," *Analog Science Fiction and Fact* 113, no. 4 (March 1993): 100.

7. http://wikipedia.org/.

8. Marc Stiegler, *David's Sling* (New York: Baen, 1988).

9. Albert W. Kuhfeld, "Spacewar," *Analog Science Fiction and Fact* 87, no. 5 (July 1971): 67.

CHAPTER 4. LOOKING INSIDE: NEW TECHNOLOGIES AND MEDICINE

1. Howard Sochurek, "Medicine's New Vision," *National Geographic*, January 1987, p. 2. An impressively illustrated introduction to some of the major medical imaging techniques, and still a good introduction even though many refinements of the techniques have been developed since its publication.

2. There is also now a special form of MRI, called *functional MRI (fMRI)*, that can be used in this way.

3. Julie Moran Alterio, "Device to Find Melanoma in Final Test Stage," *Journal News* (White Plains, NY), December 25, 2006, p. 8. Updates may be found at http://eo-sciences.com, the Web site of Electro-Optical Sciences, Inc.

4. Anthony Infantolino, "Ambulatory Swallowed Camera Capsule," http://www.jeffersonhospital.org/gastro/article11432.html (accessed August 16, 2007).

CHAPTER 5. COMPUTERS AND GENES

1. J. D. Watson and F. H. C. Crick, "A Structure for Deoxyribose Nucleic Acid," *Nature* 171, no. 4356 (April 2, 1953): 737; also available at http:// nature.com/nature/watsoncrick.pdf (accessed August 20, 2007).

2. James Dewey Watson, *The Double Helix: A Personal Account of the Discovery of the Structure of DNA* (New York: Atheneum, 1980 [originally published in 1968]).

3. Lynne Osman Elkin, "Rosalind Franklin and the Double Helix," *Physics Today* 56, no. 3 (February 2003): 61; also available at http://www.aip .org/pt/vol-56/iss-3/p42.html (accessed March 16, 2007).

4. http://www.ornl.gov/sci/techresources/Human_Genome/project/info .shtml#how for general information. The video can be accessed through that site, or directly at http://jgi.doe.gov/education/how/ (both accessed March 16, 2007).

CHAPTER 6. NEW DIRECTIONS IN BIOTECHNOLOGY

1. A good introduction to the history and current status of IVF is http:// en.wikipedia.org/wiki/In_vitro_fertilisation#.22In_vitro.22 (accessed March 16, 2007).

2. Jared Diamond, *Guns, Germs, and Steel* (New York: W. W. Norton, 1999), pp. 85–175. (Though I've identified these chapters as especially pertinent, the theme runs through much of the book.)

3. A historical time line showing milestones in genetic research from 1910 to the present can be found at http://jgi.doe.gov/education/timeline_2.html (accessed March 16, 2007).

4. E.g., Jonathan Bandler, "DNA Clears Peekskill, New York Man after Serving 15 Years in Prison," *Journal News*, September 20, 2006; other references can be found through http://en.wikipedia.org/wiki/Jeffrey_Mark _Deskovic. The tools for such forensic work were available and were used, but the court ignored them in favor of a confession obtained under duress. The

reversal of this conviction hinged not upon new examination of the DNA evidence (which the original jury knew about), but on a new confession from someone whose DNA did match. In general, judges and juries are unlikely to be able to gauge for themselves the value of new technologies as evidence, so they tend not to be generally accepted by the legal community until well after they have been recognized by scientists and engineers.

5. Deborah Levenson, "Ready or Not, Here Comes Pharmacogenomics: Will FDA's Expected Change to the Warfarin Label Make Testing Mainstream?" *Clinical Laboratory News* 33, no. 3 (March 2007): 1.

6. Richard Pizzi, "The Era of Personalized Nutrition: Has the Nutrigenomics Revolution Begun?" *Clinical Laboratory News* 32, no. 7 (July 2006): 1.

7. Joan Slonczewski, "Stem Cells and Human Cloning," *Analog Science Fiction and Fact* 122, no. 11 (November 2002): 26.

8. A Google search for "population growth" and "sustainability" on August 21, 2007, produced 2,100,000 hits. Some relevant examples, each with many other references, are: Albert A. Bartlett, "Reflections on Sustainability, Population Growth, and the Environment—Revisited," *Renewable Resources Journal* 15, no. 4 (Winter 1997–1998): 6; also available at http://dieoff.org/page146.htm; David Pimentel et al., "Impact of Population Growth on Food Supplies and the Environment," presented at AAAS Annual Meeting, Baltimore, MD, February 9, 1996; "Sustainability," http://en.wikipedia.org/wiki/Sustainability.

9. People have been arguing about the ethical problems of cloning and "designer babies" for some decades, but for most of them, until recently, it has been a rather academic "what-if" game. Now it's on the threshold of becoming an everyday, real-life concern.

10. Theodore Sturgeon, "When You Care, When You Love," *Magazine of Fantasy and Science Fiction*, September 1962; also in Robert Silverberg, *Arbor House Treasury of Modern Science Fiction* (New York: Arbor House, 1980), p. 145.

11. Ursula K. Le Guin, "Nine Lives," *Playboy*, November 1969.

12. Michael F. Flynn, "The Adventure of the Laughing Clone," *Analog Science Fiction and Fact* 108, no. 10 (October 1988): 52.

13. John Clute and Peter Nicholls, eds., *The Encyclopedia of Science Fiction* (New York: St. Martin's Griffin, 1995), pp. 236–37.

14. It's also true that a created clone will be much younger than the prototype, but they are still *genetically* identical.

CHAPTER 7. COGNITIVE SCIENCE: HOW DO WE KNOW?

1. Rather like that described in Leo Vernon, "Tools for Brains," *Astounding Science-Fiction* 23, no. 5 (July 1939): 122.

2. The one I used most often fitted a kind of curve called a Lorentzian to data from Mössbauer effect experiments—and that's *all* it did.

3. An overview of the development of graphical user interfaces can be found at http://www.sensomatic.com/chz/gui/history.html (accessed June 17, 2007).

4. Giacomo Rizzolatti, Leonardo Fogassi, and Vittorio Gallese, "Mirrors in the Mind," *Scientific American*, November 2006, p. 54. See also Vilayanur S. Ramachandran and Lindsay M. Oberman, "Broken Mirrors: A Theory of Autism," *Scientific American*, November 2006, p. 63.

5. Isaac Asimov, *Foundation* (New York: Gnome Press, 1951); *Foundation and Empire* (New York: Gnome Press, 1952); *Second Foundation* (New York: Gnome Press, 1953). All three, like most of Asimov's works, have often been reprinted by various publishers.

6. Richard Dawkins, *The Selfish Gene* (New York: Oxford University Press, 1976), p. 206.

7. See, for example: H. Keith Henson, "Memetics and the Modular Mind: Modeling the Development of Social Movements," *Analog Science Fiction and Fact* 107, no. 8 (August 1987): 291; Stanley Schmidt, "The Memetic Menace," *Analog Science Fiction and Fact* 107, no. 8 (August 1987): 4; Gary W. Strong and William Sims Bainbridge, "Memetics: A Potential New Science," in *Converging Technologies for Improving Human Performance*, ed. Mihail C. Roco and William Sims Bainbridge (Arlington, VA: National Science Foundation, 2002), p. 279.

8. Alan Turing, "Computing Machinery and Intelligence," *Mind: A Quarterly Review of Psychology and Philosophy* 59, no. 236 (October 1950): 433.

9. Isaac Asimov, *I, Robot* (New York: Gnome Press, 1950).

10. E.g., Joan Slonczewski, *The Children Star* (New York: Tor, 1999).

11. Rick Cook, "Neural Nets," *Analog Science Fiction and Fact* 109, no. 8 (August 1989): 86.

12. Bart Kosko, *Fuzzy Thinking: The New Science of Fuzzy Logic* (New York: Hyperion, 1993).

CHAPTER 8. THE EXPLOSION IN INFORMATION TECHNOLOGY

1. Konstantin Tsiolkovskii, "The Exploration of Cosmic Space by Means of Reaction Devices," *Nauchnoye Obozrnye (Scientific Review)* 10, no. 5 (May 1903) (in Russian).

2. Isaac Asimov, *Isaac Asimov's Biographical Encyclopedia of Science and Technology*, new rev. ed. (New York: Doubleday, 1972), p. 601; Anonymous editorial, "A Severe Strain on Credulity," *New York Times*, January 13, 1920, p. 12.

3. Arthur C. Clarke, "Extra-terrestrial Relays: Can Rocket Stations Give World-Wide Radio Coverage?" *Wireless World* (October 1945): 305.

4. http://en.wikipedia.org/wiki/John_R._Pierce (accessed August 22, 2007).

5. Albert Einstein, "Zur Quantentheorie der Strahlung," *Physikalische Zeitschrift* 18 (1917): 121; English translation, "On the Quantum Theory of Radiation," in *The Old Quantum Theory*, ed. D. ter Haar (New York: Pergamon, 1967), p. 167.

6. C. H. Townes and A. L. Schawlow, "Infrared and Optical Masers," *Physical Review* 112, no. 6 (December 1958): 1940.

7. http://en.wikipedia.org/wiki/Laser (accessed August 22, 2007).

8. Some people have set up "sharing" Web sites where no downloading fee is required, but others are operated as businesses (the new form of "stock agency") and incorporate security software to prevent downloading without payment.

9. Paul Levinson, *Cellphone: The Story of the World's Most Mobile Medium, and How It Has Transformed Everything* (New York: St. Martin's Griffin, 1995).

10. Eric S. Raymond, "Open Minds, Open Source," *Analog Science Fiction and Fact* 124, nos. 7, 8 (July–August 2004): 100.

11. http://en.wikipedia.org/.

12. Roy Rosenzweig, "Can History Be Open Source? *Wikipedia* and the Future of the Past," *Journal of American History* 93, no. 1 (2006): 117. Available online at http://chnm.gmu.edu/resources/essays/d/42 (accessed August 18, 2007).

13. See, for example, http://www.poynter.org/dg/lts.id.2/aid/1766/column .htm (accessed August 20, 2007).

14. Edward M. Lerner, "Beyond This Point Be RFIDs," *Analog Science Fiction and Fact* 127, no. 9 (September 2007): 44. For a fictional treatment of

some of the possible implications, by the same author, see also "The Day of the RFIDs," in the collection *Creative Destruction* (Wildside Press, 2006), pp. 15–36. (This reference, by the way, illustrates another growing trend: the book does not give a geographical address for the publisher, but only the URL www.wildsidepress.com.)

15. See, for example, "'Are We Losing Our Memory?' or The Museum of Lost Technology," http://lostmag.com/issue3/memory.php (accessed August 20, 2007; originally published in *Lost Magazine* [February 2006]).

16. Harry A. Atwater, "Plasmonics," *Scientific American*, April 2007, p. 56.

CHAPTER 9. NANOTECHNOLOGY

1. Richard Feynman, "There's Plenty of Room at the Bottom," talk at annual meeting of the American Physical Society, December 29, 1959. Reprinted in *Miniaturization*, ed. H. D. Gilbert (New York: Reinhold, 1961); available online at http://www.its.caltech.edu/~feynman/plenty.html (accessed August 18, 2007).

2. K. Eric Drexler, "Molecular Engineering: An Approach to the Development of General Capabilities for Molecular Manipulation," *Proceedings of the National Academy of Sciences USA* 78, no. 9 (September 1981): 5275. The ideas in this technical paper were subsequently developed in more depth, but at a less technical level, in the same author's book *Engines of Creation: The Coming Era of Nanotechnology* (New York: Anchor/Doubleday, 1986); and later, again at a more technical level, in his *Nanosystems: Molecular Machinery, Manufacturing, and Computation* (New York: Wiley, 1992).

3. The Nanofactory Collaboration Web site is http://www.Molecular Assembler.com/Nanofactory.

4. J. Storrs Hall, *Nanofuture: What's Next for Nanotechnology* (Amherst, NY: Prometheus Books, 2005), p. 99. Hall bases the engine results on earlier discussion in K. Eric Drexler, *Nanosystems: Molecular Machinery, Manufacturing, and Computation* (New York: Wiley, 1992), pp. 397–98.

5. Stephen L. Gillett, "Near-Term Nanotechnology," *Analog Science Fiction and Fact* 118, no. 10 (October 1998): 26. Also, by the same author, "Pollution, Solutions, Elution, and Nanotechnology," *Analog Science Fiction and Fact* 126, no. 1 (2006): 28; and "Toward a Not-Just-Diamond Age," *Analog Science Fiction and Fact* 127, no. 3 (March 2007): 54.

6. Hall, *Nanofuture*, p. 54.

7. Ibid., p. 51.

8. https://www.zurich.ibm.com/news/04.stm.html (accessed August 20, 2007).

9. Hall, *Nanofuture*, p. 55.

10. See, for example, Robert A. Freitas Jr., "Progress in Nanomedicine and Medical Robotics," chapter 13 in *Handbook of Theoretical and Computational Nanotechnology: Volume 6 Bioinformatics, Nanomedicine, and Drug Design*, ed. Michael Rieth and Wolfram Schommers (Stevenson Ranch, CA: American Scientific Publishers, 2006), pp. 619–72.

CHAPTER 10. METACONVERGENCES: WHEN BIG STREAMS MAKE STILL BIGGER STREAMS

1. Anson MacDonald (pseudonym for Robert A. Heinlein), "Waldo," *Astounding Science-Fiction* 29, no. 6 (August 1942): 9. Reprinted in book form as *Waldo: Genius in Orbit* (New York: Avon, 1950).

2. Isaac Asimov, *The Caves of Steel* (New York: Doubleday, 1954).

3. Isaac Asimov, *The Naked Sun* (New York: Doubleday, 1958).

4. See, for example, Marvin L. Minsky, *The Society of Mind* (New York: Simon & Schuster, 1985); Hans Moravec, *Mind Children: The Future of Robot and Human Intelligence* (Cambridge, MA: Harvard University Press, 1988); and J. Storrs Hall, *Beyond AI* (Amherst, NY: Prometheus Books, 2007).

5. Isaac Asimov, *I, Robot* (New York: Gnome Press, 1950). This is a collection containing all of Asimov's early robot stories, often reprinted by various publishers; the Three Laws are usually listed somewhere in the front matter.

6. Ibid.

7. Hall, *Beyond AI*.

8. Stanley Schmidt, "Immortality for Whom?" *Analog Science Fiction and Fact* 121, no. 10 (October 2001): 4.

9. Moravec, *Mind Children*, pp. 108–12.

10. See, for example, Brenda Cooper and Larry Niven, "Finding Myself," *Analog Science Fiction and Fact* 122, no. 6 (June 2002): 116; and Vernor Vinge, "The Cookie Monster," *Analog Science Fiction and Fact* 123, no. 10 (October 2003): 8.

11. J. Storrs Hall, *Nanofuture: What's Next for Nanotechnology* (Amherst, NY: Prometheus Books, 2005).

12. Robert A. Freitas Jr., "Microbivores: Artificial Mechanical Phagocytes Using Digest and Discharge Protocol," *Journal of Evolution and Technology* 14 (April 2005): 55–106; also available at http://www.jetpress.org/volume14/ freitas.pdf.

13. Robert A. Freitas Jr., *Nanomedicine, Vol. I: Basic Capabilities* (Austin, TX: Landes Bioscience, 1999); also at http://www.nanomedicine.com/NMI.htm.

14. Robert A. Freitas Jr., "Comprehensive Nanorobotic Control of Human Morbidity and Aging," chapter 15 in *The Future of Aging: Pathways to Human Life Extension*, ed. Gregory M. Fahy, Michael D. West, L. Stephen Coles, and Steven B. Harris (New York: Springer, 2008).

15. Robert J. Sawyer, *Rollback*, serialized in *Analog Science Fiction and Fact* in four parts, beginning in 126, no. 6 (October 2006): 8.

16. Robert A. Freitas Jr., "Exploratory Design in Medical Nanotechnology: A Mechanical Artificial Red Cell," *Artificial Cells, Blood Substitutes, and Immobilization Biotechnology* 26 (1998): 411–30; also available at http://www .foresight.org/Nanomedicine/Respirocytes.html (accessed June 20, 2007).

17. Synthesizers that can use too wide a range of feedstocks are potentially dangerous, though, and likely to be banned. See, for example, Robert A. Freitas Jr., "Some Limits to Global Ecophagy by Biovorous Nanoreplicators, with Public Policy Recommendations," Zyvex Corp. report, April 2000, at http:// www.rfreitas.com/Nano/Ecophagy.htm.

18. Hall, *Nanofuture*, pp. 126–33.

19. J. Storrs Hall, "Utility Fog: The Stuff That Dreams Are Made Of," in *Nanotechnology: Research and Perspectives*, ed. B. C. Crandall (Cambridge, MA: MIT Press, 1996), pp. 161–84. A briefer, less technical description is found in Hall, *Nanofuture*, pp. 188–95.

20. Storrs, *Nanofuture*, pp. 158–66.

21. Clifford D. Simak, *City* (New York: Gnome Press, 1954). The book appeared in 1954, but portions of it had already been published as short stories during the 1940s.

22. Isaac Asimov, *The Naked Sun* (New York: Doubleday, 1957).

23. Gerard K. O'Neill, *The High Frontier* (New York: Morrow, 1977). Such colonies are sometimes called "L5 colonies" because the best places to put them are at two specific points, called L4 and L5, where their orbits would be especially stable.

24. Stephen L. Gillett, "Nanotech Rocket Fuel," *Analog Science Fiction and Fact* 127, no. 10 (October 2007): 42.

25. G. Harry Stine, *Halfway to Anywhere: Achieving America's Destiny in Space* (New York: Evans, 1996).

26. Storrs, *Nanofuture*, pp. 181–83.

27. Roger Arnold and Donald Kingsbury, "The Spaceport," *Analog Science Fiction and Fact* 99, no. 11 (November 1979): 48, continued in *Analog Science Fiction and Fact* 99, no. 12 (December 1979): 61.

28. Storrs, *Nanofuture*, pp. 176–78.

29. Schmidt, *Argonaut*.

CHAPTER 11. POTENTIALS AND PROMISES

1. Replacing lungs with gills would make these people unable to live on land without technological aids, essentially reversing our present situation. Some might still choose to do it, as a means of expanding the living space available to our kind. Of course, philosophical and legal questions would then arise about how much modification people could undergo and still be considered "our kind."

2. Barry B. Longyear, "The Good Kill," *Analog Science Fiction and Fact* 126, no. 11 (November 2006): 8.

3. Robert A. Freitas Jr., "The Future of Computers," *Analog Science Fiction and Fact* 116, no. 3 (March 1996): 57–73.

4. Use would be "casual" in the sense that people could make what they wanted when they wanted it, without concern for costs and benefits. I distinguish it from "consumption" because the resources (except energy) would not be used up, but reused repeatedly.

5. Frederick Jackson Turner, "The Significance of the Frontier in American History," talk at the World's Columbian Exposition, Chicago, July 12, 1893. Turner argued that the availability of a frontier without obvious limits played a major role in shaping the American character. Of course, we now know that "limitless" really just means that there are too few of us to push the limits— but that tends to change with time. People once viewed the oceans as limitless, too, both as a food source and as a place to dump waste.

CHAPTER 12. PITFALLS AND PERILS

1. J. Storrs Hall, *Nanofuture: What's Next for Nanotechnology* (Amherst, NY: Prometheus Books, 2005), pp. 215–19.

2. Robert A. Freitas Jr. and Ralph C. Merkle, *Kinematic Self-Replicating Machines* (Georgetown, TX: Landes Bioscience, 2004), section 6.3.1; Hall, *Nanofuture*, p. 229; Robert A. Freitas Jr., "Molecular Manufacturing: Too Dangerous to Allow?" *Nanotechnology Perceptions: A Review of Ultraprecision Engineering and Nanotechnology* 2 (March 2006): 15–24.

3. Greg Bear, *Blood Music* (New York: Arbor House, 1985).

4. Wil McCarthy, *Bloom* (New York: Del Rey, 1998).

5. Probably the first, and surely the best known, attempt to estimate the probability of other technological civilizations is known as the Drake Equation, after the American radio astronomer Frank Drake. A good discussion of the method and results can be found in I. S. Shklovskii and Carl Sagan, *Intelligent Life in the Universe* (San Francisco: Holden-Day, 1966), pp. 409–18. The method involves making science-based estimates for factors such as the mean rate of star formation, the fraction of stars that have planets, the fraction of planets that develop life, and the average lifetime of a technological civilization. There's clearly much room for uncertainty in these estimates. Different authors have used different values, and the most widely accepted values have changed in the light of new knowledge, such as the recent discoveries of many extrasolar planets. But virtually everyone who considers the question uses some version of the Drake Equation.

6. See, for example, David Brin, "Xenology: The New Science of Asking, 'Who's Out There?'" *Analog Science Fiction and Fact* 103, no. 5 (May 1983): 64; and, by the same author, "Just How Dangerous Is the Galaxy?" *Analog Science Fiction and Fact* 105, no. 7 (July 1985): 80.

7. See, for example, Robert Zubrin, "Galactic Society," *Analog Science Fiction and Fact* 122, no. 4 (April 2002): 28; Ben Bova, "Isaac Was Right: N Equals One," *Analog Science Fiction and Fact* 123, no. 4 (April 2003): 40; and Stanley Schmidt, "Still Guessing, after All These Years," *Analog Science Fiction and Fact* 123, no. 4 (April 2003): 4.

8. Stanley Schmidt, "The Fermi Plague," *Analog Science Fiction and Fact* 118, no. 10 (October 1998): 4.

9. C. Northcote Parkinson, *Parkinson's Law: The Pursuit of Progress* (London: John Murray, 1957).

10. Stanley Schmidt, "The Ironic Epidemic," *Analog Science Fiction and Fact* 113, no. 14 (December 1993): 4.

11. Stanley Schmidt, "Haste Makes Haste," *Analog Science Fiction and Fact* 120, no. 1 (January 1999): 4.

12. You may wonder why withdrawal into a virtual world should matter, if

a fully automatic infrastructure including synthesizers takes care of all basic human needs. First, that kind of world, requiring no human attention to run it, represents an extreme development of nanotechnology, and the "withdrawal" problem could become severe long before we reached that point. Second, such a world might work fine as long as nothing happened to disturb its stable equilibrium. But it would be very vulnerable if, for example, another country that had not retreated into virtuality decided to invade and take over the resources (including real estate) of one that had done so.

13. In the essay "Hazards of Prophecy: The Failure of Imagination," in Arthur C. Clarke, *Profiles of the Future: An Inquiry into the Limits of the Possible* (New York: Harper & Row, 1962).

14. Stanley Schmidt, "The Dark Side of Clarke's Law," *Analog Science Fiction and Fact* 96, no. 13 (November 1996): 4.

15. Rajnar Vajra, "Emerald River, Pearl Sky," *Analog Science Fiction and Fact* 127, nos. 1, 2 (January–February 2007): 9.

16. Stanley Schmidt, "Achilles' Grid," *Analog Science Fiction and Fact* 124, no. 3 (March 2004): 4.

17. James Burke, *Connections* (London: MacMillan, 1978), p. 12. *Connections* is a fascinating study of the interconnectedness of developments in the history of technology, with more emphasis on the past and less on the future than this book.

CHAPTER 13. GETTING THERE FROM HERE: CHALLENGES AND STRATEGIES

1. It's true that some people (myself included) are fortunate enough to have work they find so intrinsically satisfying that they don't think of it as "just a job." But in these cases the work really has two parts: something-that-I-like-to-do, and something-I-must-do-to-make-a-living. The hidden assumption made almost universally in our society is that the two must go together. Future societies may find ways to decouple them, enabling people to do more of what they find most satisfying while feeling less need to do specific jobs to keep food on the table.

2. I mentioned earlier that any aspiring Big Brother would now have access to tools of oppression that George Orwell never imagined. Here is a case where the opposition could also gain a few advantages!

3. David Brin, *The Transparent Society: Will Technology Force Us to Choose between Privacy and Freedom?* (Reading, MA: Addison-Wesley, 1998).

4. Robert J. Sawyer, *Hominids* (New York: Tor, 2002).

5. Robert J. Sawyer, *Humans* (New York: Tor, 2003).

6. Robert J. Sawyer, *Hybrids* (New York: Tor, 2003).

7. Robert A. Freitas Jr., "What Price Freedom?" *Nanotechnology Perceptions: A Review of Ultraprecision Engineering and Technology* 2 (May 2006): 99–106; also available at http://www.rfreitas.com/Nano/WhatPriceFreedom.pdf.

8. Stanley Schmidt, "A Flash of Darkness," *Analog Science Fiction and Fact* 82, no. 1 (September 1968): 145.

9. J. Storrs Hall, *Nanofuture: What's Next for Nanotechnology* (Amherst, NY: Prometheus Books, 2005), pp. 220–25.

10. Faculty members of several New England academic institutes have established the New England Complex Systems Institute, dedicated to the study of complex systems and hosting periodic conferences at which many fields converge. I participated in their International Conference on Complex Systems in Boston in June 2006, and was impressed by the diversity of both topics and speakers, many of them discussing topics extending well beyond their "official" specialties.

INDEX

263